Malcolm William Hilles, Henry Savage

The Pocket Anatomist

Second Edition

Malcolm William Hilles, Henry Savage

The Pocket Anatomist
Second Edition

ISBN/EAN: 9783337365486

Printed in Europe, USA, Canada, Australia, Japan

Cover: Foto ©berggeist007 / pixelio.de

More available books at **www.hansebooks.com**

THE
POCKET ANATOMIST:

BEING

A COMPLETE DESCRIPTION OF THE ANATOMY OF THE HUMAN BODY,

FOR

THE USE OF STUDENTS,

BY

M. W. HILLES,

FORMERLY LECTURER ON ANATOMY AND PHYSIOLOGY AT THE WESTMINSTER HOSPITAL SCHOOL OF MEDICINE, ETC.

SECOND EDITION.

PHILADELPHIA:
LINDSAY AND BLAKISTON.
1867.

PREFACE.

To the student preparing for his examination at the several Medical Colleges, works which contain a condensed description of the structures and functions of the human body are essentially useful, as they not only convey concise information on these subjects, but they also direct his attention to the most important points connected therewith. Such works are not intended to supersede the more elaborate treatises on the various branches of medical science: these must be studied also, as, without their aid, it is vain to suppose, that sufficient information can be obtained to enable the student to enter upon the pursuit of the profession which he has selected, with credit to himself, or with advantage to those who may intrust their health and lives to his care.

There is no "short road to knowledge:" the foundation must be first carefully laid, by close study and diligent research; when this has been accomplished, the attention will be profitably directed to the arrangement, classification, and condensation of the knowledge thus obtained: knowledge is not a burden, nor is it unprofitable: it is only in the pursuits of after life that we feel the truth of this, and begin to experience the advantage of having laid up, in our earlier years, a store of information upon which we may draw fearlessly on occasions of difficulty.

Anatomy may be said to be the basis on which all medical science rests; without correct and extensive anatomical knowledge the practice of surgery must be fraught with danger, and that of medicine empirical and of doubtful advantage. The medical bodies of the United Kingdom happily insist on all candidates for their degrees or diplomas possessing a competent knowledge of anatomy; it is much to be regretted that a deficient supply of subjects should still exist, and present serious obstacles to the attainment of that practical information which is so desirable.

This work is based on a treatise favourably known to the profession as *Savage's Anatomist*, and written by my friend and former colleague, Mr. Henry Savage. It now appears, from its enlarged and altered form, rather as a new work than a new edition of a former treatise. I trust that in this form it will continue to meet with approval and support.

CONTENTS.

CHAPTER I.

	PAGE
THE HEAD	13
Cranium	13
Occipito frontalis	13
Muscles of External Ear	14
Face	14
Muscles of ditto	14
" nose	16
" lower jaw	16
Vessels and nerves of face	16
Parotid gland	17

CHAPTER II.

THE NECK	19
Muscles of ditto	19
" between lower jaw and os hyoides	20
" of tongue	20
" between os hyoides and sternum	21
Thyroid gland	22
Triangles of neck	23
Submaxillary gland	24
Sublingual ditto	24
Vessels and nerves of neck	25
Deep muscles	25
Larynx	26
Thyroid cartilage	27
Cricoid "	27
Arytænoid "	27

	PAGE
THE NECK	
Epiglottis	28
Epiglottic gland	28
Muscles of larynx	29
Pharynx	30
Muscles of ditto	31
Arteries and nerves	32
Velum	32
Tonsil	33

CHAPTER III.

THE UPPER EXTREMITIES	35
Muscles of anterior and lateral parts of trunk	35
Muscles on posterior part of trunk	37
Superior extremity	41
Muscles of shoulder	41
" of arm	43
Arteries of shoulder	44
Nerves of ditto	45
Axilla	45
Mammary gland	45
Fore arm	46
Muscles of ditto	47
Vessels and nerves of ditto	50
Hand	51
Muscles of ditto	52
Vessels and nerves of ditto	54

CHAPTER IV.

THE TRUNK	55
Thorax	55
Pleura	56
Anterior Mediastinum	56
Thymus gland	57
Lungs	57
Bronchial tubes	58
Vessels and nerves of lungs	58
Trachea	59
Posterior Mediastinum	60

CONTENTS.

The Trunk.

	PAGE
Œsophagus	60
Vena azygos	60
Thoracic duct	61
Abdomen	62
Muscles of abdomen	62
Diaphragm	65
Viscera of abdomen	67
Peritoneum	67
Stomach	69
Small intestines	71
Large ditto	72
Liver	74
Gall bladder	76
Pancreas	77
Spleen	77
Hernia	78
Inguinal ditto	78
Femoral ditto	92
Urinary organs	100
Kidneys	100
Ureters	101
Suprarenal capsules	101
Bladder	102
Urethra	103
Male organs of generation	104
Scrotum	104
Testes	104
Spermatic cord	106
Vesiculæ seminales	106
Prostrate gland	106
Penis	107
Perineum	108
Muscles of	108
Female organs of generation	111
Uterus	112
Ovaries	113

CHAPTER V.

The Inferior or Lower Extremities	114
Muscles of	114

CONTENTS.

	PAGE
THE INFERIOR OR LOWER EXTREMITIES.	
Muscles of hip joint	114
" front and sides of thigh	117
" posterior part of thigh	119
Vessels and nerves of thigh	120
Leg	121
Muscles of	121
Foot	124
Muscles of	124
Vessels and nerves of leg and foot	127

CHAPTER VI.

FASCIÆ	129
Cervical	130
Abdominal	130
Pelvic	132
Perineal	132
Lata	133
Crural	135
Plantar	136
Brachial	137
Palmar	138

CHAPTER VII.

VASCULAR SYSTEM	139
Heart	139
" structure of	142
" development of	143
" investments of	143
Arteries	145
pulmonary	145
aorta	145
innominata	147
carotid	147
" right common	147
" left common	147
" External	148
anterior branches	148

CONTENTS.

VASCULAR SYSTEM.

	PAGE
Arteries posterior	149
inner	149
terminal	149
internal	150
branches	150
Cerebral branches from Basilar	151
Subclavian	152
" right	152
" left	152
branches	153
Axillary	154
branches	154
Brachial	155
branches	155
Radial	156
branches	156
Ulnar	157
branches	157
Aorta	
Thoracic	158
Abdominal	159
Cœliac axis	159
Hepatic	160
Splenic	160
Sup. Mesenteric	160
Inf. Mesenteric	161
Renal	161
Spermatic	162
Terminal branches	162
Iliacæ Communes	162
" internal	163
branches	163
Gluteal	164
Ischiatic	164
Int. Pudic	165
" External	166
branches	166
Femoral	166
branches	167
Profunda	167

VASCULAR SYSTEM— PAGE
  AORTA—

      Popliteal... 168
         branches.. 168
      Tibial anterior.................................... 168
        " posterior................................... 169
         branches....................................... 169
Collateral circulation.................................. 170
Venous System.. 172
Vein, Internal Jugular.................................. 172
  " Labial... 172
  " Ext. Jugular..................................... 172
  " Axillary... 172
Vena Innom., right..................................... 173
  " left... 173
Vena cava, superior..................................... 173
  " Vein, int. saphena............................. 173
  " ext. saphena.................................. 173
  " popliteal...................................... 173
  " common iliac................................. 173
Vena cava inferior...................................... 174
  " porta.. 174
Lymphatic system....................................... 174
Thoracic duct... 174

CHAPTER VIII.

THE NERVOUS SYSTEM................................... 176
  Spinal cord... 176
  Medula oblongata.................................... 179
  Corpus olivaria...................................... 180
    " restiformia.................................... 180
  Filamenta arciformia................................ 180
  Pons Varolii... 181
  Crura cerebri.. 181
  Cerebellum... 183
  Cerebrum... 185
  Tuber cinereum...................................... 186
  Fissure of Sylvius................................... 187
  Corpus callosum..................................... 188
    " striatum 189

THE NERVOUS SYSTEM.

	PAGE
Third Ventricle	190
Membranes of brain	191
Dura mater	192
Sinuses	192
Arachnoid	194
Pia mater	194
Structure of the brain	194
Vessels	199
Spinal nerves	200
Cervical plexus	203
Brachial	204
Intercostal nerves	206
Lumbar plexus	207
Sacral	209
Cerebral nerves	211
Sympathetic system	214
Cerebral ganglia	216
Semilunar ganglion	216

CHAPTER IX.

BONES, LIGAMENTS, AND JOINTS	218
Bones of the head	218
frontal	218
parietal	219
occipital	220
temporal	221
sphenoid	223
ethmoid	224
of face	225
Inferior maxilla	228
Temporo-maxillary articulation	229
Spine	230
" ligaments	231
Ribs	232
Sternum	233
Ligaments of ribs	234
Bones of upper extremity	234
Shoulder joint	237
Bones of fore-arm	237

CONTENTS.

	PAGE
BONES OF THE HEAD	
Elbow-joint	238
Carpus	239
Wrist-joint	240
Bones of inferior extremity	241
pelvis	241
" articulations of	244
femur	245
Hip-joint	246
Knee-joint	248
Ankle-joint	252

CHAPTER X.

ORGANS OF SENSE	254
Eye	254
Nose	259
Ear	260
Tongue	262
Skin	263

THE ANATOMIST.

The head, neck, trunk, and extremities form the principal divisions of the human body.

CHAPTER I.

THE HEAD

Consists of the Cranium and Face.

SECTION I.

The Cranium is covered by the scalp, composed of—1. The Skin. 2. The occipito-frontalis muscle. 3. The periosteum.

The skin of the scalp is dense and firm, and is furnished with hair and sebaceous follicles and granulated fat.

Occipito-frontalis is bicipital, with a central tendon or aponeurosis, *Or.* two ext. thirds of the super. transverse occipital ridge, and adjoining part of the mastoid process; *Ins.* nasal bones, int. angular process, the rest of its fibres intermixing with the corrugator supercilii, and orbicularis palp. Covers the whole skull, excepting the temporal fossæ. *Use*, to raise the skin of the forehead, eyebrows, and eyelid. The scalp is supplied with blood from the frontal arteries in front the temporal and aural on the sides, and the occipital posteriorly; its nerves are derived from the fifth, portio dura, ascending cervical, and first cervical branches; it is loosely connected to the periosteum and bone by areolar tissue, and thus moves freely on the skull. On the side of the head are

The Muscles of the External Ear, 3.

Superior auris, or *attollens aurem*, Or. from the cranial aponeurosis on the side of the cranium, immediately above the ext. ear; *Ins.* the upper and anterior part of the cartilage of the ear.

Anterior auris, or *attrahens aurem*, Or. posterior part of zygomatic process, and cranial aponeurosis; *Ins.* front part of helix.

Posterior auris, or *retrahens aurum*, Or. from the the mastoid process above sterno-mastoid; *Ins.* back part of concha. The *use* of these muscles is indicated by their names.

Section II.

The Face

is covered by a delicate vascular skin, closely connected with the subjacent

Muscles, 33.

Orbicularis palpebrarum, Or. the upper edge of the tendo oculi, and int. ang. process; *Ins.* the lower edge of the tendon, ascending process of the maxilla sup., and inner third of the lower edge of the orbit. The tendon crosses the junction of the lower two-thirds with the upper third of the lacrymal sac. *Use*, to close the eyelids, by depressing the upper, to compress the lacrymal sac, and force the tears into the nasal duct.

Corrugator supercilii, Or. inner end of the superciliary ridge; *Ins* orbic. palpebr. at its middle. *The following four m. arise in succession from the inner canth.*

Levator labii super. alæque nasi, Or. narrow from the inner angular process beneath the tendo palpebræ; *Ins.* skin of the ala nasi and upper lip, and into the orbic. oris muscle.

Levator labii proprius, Or. broad, inner half of the edge of the orbit; *Ins.* skin and hair bulbs of the upper lip; supposed to render the hair erect, as in the feline tribes.

Zygomaticus minor, Or., narrow from the inf. edge of the os malæ; *Ins.* confounded with those of the above muscle.

Zygomaticus, major, Or., broad from a groove on the os mala, above its lower edge; *Ins.* narrow into the labial commissure.

Levator anguli oris (Caninus,) Or. broad from the canine fossa; *Ins.* narrow with the zyg. maj. into the labial commissure; the infra orbitar nerve and artery appear between this and the levat. prop. m.

Depressor anguli oris, Or. broad from the ext. oblique line of infer. maxilla; *Ins.* narrow with the zyg. major and caninus m. into the labial commissure. Its fibres are partly confounded with those of the platysma.

Depressor labii inf., Or. broad from the same line as, but internal to, the last m.; *Ins.* broad into the skin of the lower lip.: united with its fellow above, but separated below by the

Levator labii inf. vel menti, conical, *Or.* from a fossa close to the symphysis menti; *Ins.* by diverging fibres into the skin and hair bulbs of the skin. The *use* of the preceding muscles is indicated by their names, the zygomatici draw the angle of the mouth upwards and backwards.

Buccinator Or. 1, super. alveolæ, from that of the first molar tooth to the last; 2, corresponding alveolæ of the lower jaw; 3, ptergo maxillary ligament; *Ins.* the labial commissure, its lower fibres crossing the upper. These muscles are antagonized by the succeeding. *Use*, to press the cheek against the teeth, so as to prevent the food passing between them, to assist in mastication, articulation, blowing, and sucking.

Orbicularis oris. This sphinctic muscle is composed of two portions, corresponding to the upper and lower lips. Their fibres decussate at the labial commissure to become continuous, the upper portion with the lower fibres of the buccinator, and vice versa. Use, to close and corrugate the mouth.

Muscles of the Nose, 3—4.

Pyramidalis nasi. A few fibres of the frontalis m., descending on each side of, and parallel to, the nasal suture to be lost in the compressor nasi.

Compressor nasi, Or. narrow from the edge of the canine fossa; the two m. expand so as to cover the nasal cartilage and some part of the alæ, and unite by means of a thin fascia on the dorsum nasi.

Depressor alæ nasi, Or. narrow close to the median line from a fossa on the alveola of the incisor tooth; *Ins.* by diverging fibres into the the orbic. oris, alæ nasi, and septum.

Naso labialis, (Albinus,) from the under part of the septum nasi runs downwards and backwards, and is lost in the orbic. oris.

Muscles acting on the Lower Jaw.
To shut it, 2.

Masseter, Or. 1, by a strong fascia from the ant. two-thirds of the edge of the zygoma, and adjoining portion of the super. maxilla; 2, by a thinner fascia from the post. third of the zygoma; 3, from the inner surface of the zygoma. *Ins.* the first portion, angle of the jaw; the second, ext. surface of the ramus; and third, ext. surface of the coronoid process.

Temporalis, Or. 1, Temporal fossa; 2. Temporal aponeurosis. *Ins.* by a strong narrow tendon into the coronoid process of the inf. maxilla, and sometimes by an additional fasciculus and tendon into the inner edge of the ramus. The *horizontal motion of the lower jaw is effected by* 2 m.

Pterygoideus ext., Or. 1, the whole outer surface of the ext. pterygoid plate and adjoining portion of the tuberosity of the palate bone: 2, from a ridge on the temporal bone, which separates the temporal from the zygomatic fossa. *Ins.* fossa on the anterior surface of the neck of the condyle of the lower jaw and edge of the interarticular cartilage; the int. max. art. enters the pterygoid fossa between its two origins.

Pterygoideus ext., Or. 1, fossa between the pterygoid plates; 2, the hamular process; 3, tuberosity of the palate bone; it passes downwards, backwards, and outwards, to be *ins.* into the inner surface of the angle of the lower jaw. The interval between the muscles below is traversed by the int. maxil. art. and vein, and the dental and gustatory nerves. The dental nerve and artery, in their descent, separate the latter muscle from the ramus of the jaw. The lateral or grinding motion of the lower jaw is effected chiefly by the pterygoid muscles, and in this the muscles of either side act alternately, and not conjointly, as is the case in most other muscles. These last four muscles, although usually included amongst those of the face, do not strictly belong to this part.

In the dissection of the face, numerous vessels and nerves are found supplying the skin, muscles, &c. The principal arteries are derived from the labial branch of the ext. carotid, the frontal branch of the ophthalmic, the infra orbital, and the inferior dental branches of the internal maxillary artery; the nerves are the frontal, infra orbital, and inferior maxillary, or dental of the 5th, branches of the portio dura and of the cervical plexus. The masticatory actions of the masseter, temporal, pterygoid, and buccinator are supplied by the non ganglionic portion of the 3d division of the 5th.

The Parotid Gland

is met with in dissecting the deep muscles of the face. There are three salivary glands on each side: 1 parotid; 2 submaxillary; 3 sublingual. Of these, the parotid is the largest; it is lodged in a quadrilateral fossa behind the ramus of the lower jaw, which bounds it in front; behind, is the mastoid process; above the zygomatic arch; and below, the stylo-maxillary ligament. Its superficial surface is covered by the skin, a few fibres of the platysma, and a continuation of the cervical fascia; its deep surface passes in between the pterygoid muscles, and is in close contact with the int. carotid artery, int. jugular vein, 3rd division of the fifth nerve, the 8th and 9th cerebral nerves, and the styloid process; in its substance are found the portio dura, or pes anserinus, the ext. jugular vein, and the ext. carotid artery, and its terminating branches. This belongs to the *Conglomerate* class of glands; it is granular in structure, the granules being separated by prolongations of the cervical fascia; a portion prolonged on the masseter muscle is called the *socia parotidis*. Its duct, *Steno's duct*, arises from the granules by small radicles, which unite and form the trunk; this crosses the masseter muscle from the anterior edge of the gland, perforates the buccinator, and opens into the mouth opposite the second molar tooth of the upper jaw. A line drawn from the lobe of the ear to nearly between the ala nasi and angle of the mouth indicates its course. The structure of this duct is fibro-mucous; the fibrous or ext. coat is remarkably strong, the canal is small, particularly at its orifice in the mouth. The gland is supplied with blood from the ext. carotid branches, its nerves are derived from the cervical plexus, the portio dura, fifth, and sympathetic.

CHAPTER II.

THE NECK

includes the space between the lower jaw above, the clavicles beneath, and the cervical vertebræ posteriorly; it comprises the neck, properly so called, the mouth, pharynx and larynx, with numerous vessels and nerves.

The integuments of the neck are thin; beneath them may be noticed the projections of the principal muscles—the larynx on the mesial line, and the ext. jugular vein laterally.

Section I.

Muscles of the Neck.

Platysma myoides.—Commences in the subcutaneous tissue over the pectoralis major and deltoid; *Ins.* into the fascia covering the parotid and masseter m., the labial commissure, and ext. oblique line of the lower jaw; some fibres are mixed with those of the depr. angul. oris, quadratus menti, and fellow. *Use*, to depress the angle of the mouth and lower jaw, and to corrugate the skin of the neck.

Sterno-cleido-mastoideus, Or. fleshy from the sternal third or more of the clavicle, and by a round tendon, from the adjoining surface of the first bone of the sternum; *Ins.* mastoid process and sup. transverse ridge of the occipital bone as far as the trapezius. Traversed by the spinal accessory nerve; the ext. jugular vein descends some way upon it; its ant. edge is a guide to the common carotid artery. *Use*, if both muscles act, the interior portion will depress the head anteriorly, the posterior fibres posteriorly; the sternal portion of each muscle will also rotate the head to the opposite side, the clavicular portion will depress it laterally.

Between the Lower Jaw and Os Hyoides.
Depressors of the Jaw. 3

Digastricus, *Or. post. belly*, groove behind the mastoid process; *ant. belly*, a small fossa close to the symphysis menti. The two bellies are united by a round tendon, which is attached to the upper edge of the body of the os hyoides by means of a few tendinous fibres. The post belly passes through the stylo hyoid m.

Mylo hyoideus, *Or.* whole length of the mylo hyoid ridge; *Ins.* the whole length of the body of the os hyoides behind the digastric tendon, and to a median tendinous line extending from that bone to the symphysis menti. It forms the floor of the mouth.

Genio hyoideus, *Or.* the infer. of the two tubercles, on the inside of the jaw close to the symphysis; *Ins.* narrow, body of the os hyoides close to the median line.

These three muscles being removed, the following are brought into view:—

Of the Tongue, 5.

Genio hyo-glossus, *Or.* narrow tendinous, from one of the sup. tubercles on the inside of the jaw, close to the symphysis, above the former m., it expands like a fan, to be *Ins.* 1, into the body of the os hyoides; 2, the pharynx between that bone and the stylo-glossus; 3, with its fellow into the under surface of the tongue from the base to the apex; the hypoglossal nerve passes between its pharyngeal and lingual portions. *Use* to depress the tongue in the centre; the posterior fibres will draw the base of the tongue forward, the anterior will retract the tip.

Lingualis, or *Sublingualis*, similar fibres are seen on the dorsum of the tongue, is a fasciculus of fibres running from base to apex, between the above m. and the stylo and hyo glossi. *Use*, to draw the tip towards the base of the tongue, so as to render this convex superiorly.

Hyo glossus, *Or.* 1, the body of the os hyoides, near the larger cornu; 2, the whole length of this cornu. *Ins.* side of the tongue, between the lingualis and stylo glossus. The lingual artery crosses beneath it a little above the cornu, and the lingual nerve in front of it, a little above the artery. The gustatory crosses its upper margin. *Use* to depress the sides of the tongue, and render the dorsum convex.

Stylo glossus, *Or.* narrow, inf. half of the styloid process; *Ins.* side of the tongue, from the palato-glossus to the tip; a second portion divides the fibres of the hyo glossus, to join the transverse fibres of the tongue. Covered by the gustatory nerve, and sublingual gland. *Use*, to draw the tongue backwards, and to elevate the lip. *The two other styloid muscles elevate and dilate the pharynx.*

Stylo hyoideus, *Or.* narrow, back part of the styloid process, near its apex; *Ins.* body of the os hyoides, near the middle, gives passage to the post. belly of the digastricus. *Use*, to raise and draw back the os hyoides and tongue.

Stylo pharyngeus, *Or.* vaginal process and back part of the base of the styloid process, passes between the two sup. constrictors, with which its sup. fibres are lost; *Ins.* the edge of the ala of the thyroid cartilage, and side of the pharynx. The glossa pharyngeal nerve runs on its outer side. Separates the ext. from the int. carot. and int. jugular vein.

Between the Os Hyoides and Sternum, 4.

Sterno hyoideus, *Or.* post. surface of sternum, and cartilage of the first rib, sternal end of clavicle, and capsule of the joint; *Ins.* lower border of the *body* of the os hyoides. *Use*, to depress os hyoides, larynx, and pharynx.

Thyro hyoideus, Or. oblique line on the ala of the thyroid cartilage; *Ins.* post. surface of the body of the os hyoides, and part of its larger cornu. *Use*, to approximate the os hyoides and thyroid cartilage.

Sterno thyroides, Or. broad, post. surface of the sternum, below the sterno hyoideus, and from the first costal cartilage; *Ins.* oblique line of the ala of the thyroid cartilage. *Use*, to depress the larynx.

Omo hyoideus, Or. superior edge of the scapula, behind the *notch;* 2, inf. edge of the os hyoideus at the junction of its body and cornu. The two portions are united, about an inch and a half above the clavicle, by a flat tendon, which is fixed to the latter bone by a portion of the deep cervical fascia. *Use*, to draw the os hyoideus downwards, and to one side.
Two muscles extend from the sternum to the head.

Immediately beneath the preceding muscles, and lying in front of the trachea, and extending laterally on the sides of the larynx, is a soft, deep reddish body, the *thyroid gland* or *body*. This is crescentic in shape, the concavity upwards, convex anteriorly, concave posteriorly, and consists of two large lobes, the lateral lobes on either side, connected in front of, and in some rare cases behind the trachea, by a narrow isthmus, the *central lobe*. This body is larger in the infant and female subject. The structure of this body is cellular, the cells containing a thin serous fluid; it is largely supplied with blood by the inferior thyroid from the subclavian, the superior thyroid from the ext. carotid, and frequently by a middle thyroid from the arch of the aorta. Its use is wholly unknown. As is usual in such cases conjecture has been very busy on the subject.

The muscles of the neck, by their varied directions, divide the nearly square surface of this region into numerous spaces, generally of a triangular shape, which are of much importance, as containing several large bloodvessels, nerves, and other structures. The principle of these are—

1. The *Great Anterior Triangle.* Bounded above by the lower jaw and stylo maxillary lig., which form its base; in front, by the mesial line; posteriorly, by the ant. edge of the sterno mastoid; its apex is, inferiorly, at the sterno clavicular articulation.

2. The *Great Posterior Triangle.* Bounded below by the clavicle, which forms its base; in front, by the post. edge of the sterno mastoid; behind, by the ant. edge of the trapezius; its apex is superiorly, where these muscles approach, but do not meet. Each of these triangles is subdivided by the omo hyoid muscle into two as follows—

1. *Anterior Inferior Triangle.* Bounded by the mesial line as far as the os hyoideus, in front, which forms its base; above by the anter. belly of the omo hyoid; below, by the ant. edge of the sterno mastoid.; its apex is formed by the crossing of these muscles. This space contains the com. cartid art., internal jugular vein, and par vagum, with some smaller nervous filaments.

2. *Anterior Superior Triangle.* This is again subdivided into two by the digastric muscle; vix.—

1. *Infra hyoid space.* Bounded above by the os hyoides and digastric tendon, which forms its base; internally, by the ant. belly of the omo hyoid; ext. by the ant. margin of the sterno mastoid; its apex is at the decussation of these muscles. It contains the same parts, and generally the bifurcation of the com. carotid into its two branches, the ext. and int. carotids

Supra hyoid, or *digastric space.* Bounded above by the lower jaw and stylo maxillary lig., which form its base; in front, by the ant. belly of the digastric; behind, by its post. belly; its apex is at the tendon of this muscle, where it is attached to the os hyoides. This space *contains* the submaxillary and sublingual glands, and labial and lingual arteries, the gustatory, ninth, and chorda tympani nerves, and Whartonian duct.

1. *Posterior Inferior Triangle.* Bounded below by the clavicle, which forms its base; above, by the post. belly of the omo hyoid; in front, by the post. edge of the sterno mastoid; its apex is at the decussation of these muscles. It contains the subclavian vein, the subclavian art. in its third stage, and brachial plexus; and, more superficially, branches of the cervical plexus ext. jugular vein, and some lymphatic glands.

2. *Posterior Inferior Triangle.* Much larger, is bounded behind by the ant. edge of the trapezius, which forms its base; above and in front by the post. edge of the sterno mastoid; beneath, by the post. belly of the omo hyoid; its apex is at the decussation of these muscles. It contains the cervical plexus, and spinal accessory nerve, the transversalis colli art., and numerous lymphatic glands.

The submaxillary and sublingual glands met with in the digastric space form, with the parotid already described, the salivary glands.

The *submaxillary gland* is lodged in a deep fossa beneath the lower jaw; bounded above by this bone; beneath and in front by the ant. belly of the digastris; behind, by its post. belly. It is covered by the skin, platysma, and cervical fascia, and rests on the mylo hyoid and hyo glossus muscles. It is traversed by the labial art. which supplies it with blood. The structure of this gland resembles that of the parotid, but is less granular, as the cervical fascia does not send processes into its interior. Its duct, the *Whartonian duct*, is thin and delicate, and opens on the side of the frænum linguæ. A prolonged portion of this gland winds round the posterior edge of the mylo hyoid mus., and joins.

The *Sublingual gland*, the smallest of the salivary glands, lies immediately beneath the mucous memb.

of the mouth, and above the mylo hyoid musc. It opens into the mouth by numerous orifices beneath the tongue; some join the Whartonian duct.

All these parts are bound down, retained in their places, and more or less protected, by a dense layer of fibrous structure, the cervical fascia.' (See Fasciæ.)

Section II.

The Vessels and Nerves of the Neck.

The *arteries* met with in the neck are the right and left common carotid arteries, with their branches, the ext. and int. carotids, and the principal branches of them; the right and left subclavian arteries, and their principal branches; the veins are the internal and external jugular, and the subclavian, with their principal branches; the nerves are the cervical and brachial plexuses, with their principal branches; the par vagum or pneumogastric, the lingual or ninth, and sympathetic, with their numerous branches.

Section III..

Deep-seated Muscles of the Neck, 5.

The number of scaleni is variously stated. Albinus makes 5, Sabatier 3, Chaussier 1, recent writers 2.

Scalenus anticus; Or. upper surface of the first rib at its middle, *Ins* ant. transverse tubercles of the sixth, fifth, fourth, and third cervical vertebræ.·. separates the subclavian vein from the artery; the phrenic nerve descends diagonally across its ant. surface.

Scalenus posticus; Or. by two slips, 1, from the first rib between its subclavian depression and tubercle; 2. upper edge of the second, between its tubercle and angle; *Ins.* post. transverse tubercles of the six inf. cervical vertebræ·.·The subclavian artery and brachial plexus pass between the two scaleni. The use of the scaleni muscles is to elevate or fix the ribs to which they are attached, and to depress the neck forwards, and to one side.

Longus colli, consists of three distincts orders of fasciculi, 1, *Or*. ant. transverse tubercles of the fifth, fourth, and third cervical vertebræ; *Ins*. fore-part of the atlas; 2, *Or*. from the bodies of the first three dorsal vertebræ *Ins*. ant. transverse tubercles of the third and fourth cervical vert., 3. *Or*. the bodies of the four last cervical, and three sup. dorsal, and intermediate ligaments; *Ins*. front of the axis and third cervical vertebræ ·.·supports the great cervical vessel and nerves. *Use*, to bend the neck latterlly, and to rotate the atlas and head on the second vertebræ.

Rectus capitis anticus major; *Or*. ant. transverse tubercles of the third, fourth fifth, and sixth cervical vertebræ; *Ins*. broad into the basilar process just in front of the foramen magnum.·. supports the int. carot. artery, sup. cerv. gang.·.separates the ant. scaleni from the long. coli. *Use*, to bend the head and neck forward.

Rectus capitis ant. minor, beneath the former m. Extends from the base of the transverse pr. of the atlas upwards and inwards to the basilar process.·.supports the sup. cerv. ganglion. *Use*, to bend the head and neck forward and a little on one side.

Section IV.

The Larynx

is the superior extremity of the trachea, with which it is continuous. It opens superiorly into the mouth, and lies on the mesial line of the neck, having the carotid arteries, internal jugular vein, and par vagum on each side, the pharynx behind, and the sterno-hyoid, sterno-thyroid, and thyro-hyoid, muscles in front.

The larynx is composed of cartilages, fibro-cartilage, muscles, mucous membrane, and glands. Four cartilages, thyroid, cricoid, and two arytænoid; one fibro-cartilage, the epiglottis; fourteen muscles, the thyro-hyoid, crico-thyroid, thyro-arytænoid, crico-arytænoid lateralis, crico-arytænoid posticus, aryteno-epiglottic, inter. arytænoid, and thyro-epiglottic. The last two are single muscles. One gland, the epiglottic. The mucous membrane lines the interior of the larynx.

The *thyroid* cart. is the largest and superior. It consists of two quadrilat. plates, or alæ, united in front forming the pomum Adami, and separated posteriorly to receive the post ext. of the cricoid cart. The *upper* edge is irregularly convex, and is joined by the thyro hyoid lig. to the os hyoides; its *inf.* edge is nearly straight, having a slight projection about its centre, and is connected to the first ring of the trachea by an elastic tissue, the crico-thyroid ligament. The *posterior*, or free margin rounded, is prolonged superiorly into the greater or superior cornu, attached to the os hyoides by a round and sometimes cartilaginous ligament. The *inferior* is shorter, and forms an articulation on the side of the cricoid cart. The *outer* surface, nearly plane, is marked by an oblique line, to which the sterno-thyroid and thyro-hyoid muscles are attached; the *inner* surface is concave, and covers the crico-arytænoid lateralis and thyro-arytænoid muscles.

The *cricoid* cartilage is ring-shaped, and lies beneath the preceding. Narrow in front, deep behind, it fills here the interval between the alæ of the thyroid cart.; its outer surface convex, gives origin to the cricothyroid musc. and more posteriorly articulates with the inf. cornu of the thyroid cart. ; behind this the posterior surface gives origin to the crico-arytænoidei postici muscles, between which is a prominent ridge; the inner surface is lined by mucous memb., the superior edge of the cricoid cart. slopes upwards and backwards, and is marked by a convex articulating surface for the arytænoid cartilage.

The *arytænoid* cartilages are triangular in shape, the *base* is concave, and rests on the edge of the cricoid cart.; on it are two projections, one ext. for the insert. of the crico-arytænoid muscle, the other anterior for the attachment of the true chorda vocalis; the *apex* is surmounted by the appendix or corniculum. The *posterior* surface concave, lodges the inter arytænoid musc.; the anterior convex gives attachment to the false chorda vocalis, and the aryteno epiglottidean fold. The internal surface is flat, opposed to its fellow, and covered by mucous membrane. These cartilages are very moveable, and thus influence the vocal chords.

The *epiglottis* is situated immediately behind the root of the tongue, and over the upper opening of the larynx; it is the shape of an oval leaf, concave posteriorly, and covered by mucous membrane; convex in front, it is attached to the root of the tongue by three folds of mucous membrane, the frænum, the largest in the centre, the frænula on each side; to its edges are attached the aryteno epiglottidean folds; inferiorly the epiglottis ends in a stalk-like process, which is inserted into the retreating angle of the thyroid cart. above the chorda vocalis. The use of the epiglottis is to cover the glottis in the act of deglutition; it is fibro-cartilaginous in structure.

The *epiglottic gland* lies between the stalk-like process of the epiglottis and the thyro hyoid memb. Its glandular nature is very questionable.

The larynx is connected superiorly to the os hyoides by the thyro hyoid ligament; this is best marked in the centre and at each posterior margin; the thyroid and cricoid cartilages are connected in front by the crico thyroid memb. or lig.; this is strong, and of a yellowish colour; a capsular ligament connects the articulation between the cricoid cart. and the inf.

cornu of the thyroid, which is also lined by synovial membrane.

In the interior of the larynx, and forming the *rima glottidis*, are the chordæ vocales, two true and two false; the true are the stronger and inferior, the false superior and less distinct; both pass from the retreating angle of the thyroid cartilage to the tubercle on the base of the arytænoid, the false one slightly curved, the true one straight; between them is a small fossa, the sacculus laryngis; their structure is fibrous. The use of the chorda vocalis is to open and close the rima glottidis, and to affect the tones of the voice by their different states of tension.

Thyro-hyoideus, *Or.* oblique ridge of ala of thyr. cartilage; *Ins.* lower border of cornu of os hyoides. *Use*, to raise and draw forwards the larynx and close the glottis.

Crico-thyroideus, *Or.* from forepart of cricoid cart.; *Ins.* lower edge and cornu of thyroid cart. *Use*, to approximate the cartilages and make tense the vocal chords.

Thyro-arytenoideus, *Or.* inner surface of thyroid cart.; *Ins.* anterior and outer edge of arytenoid cart. *Use*, to draw the cartilages towards each other and narrow the rima glottidis.

Crico-arytenoideus posticus, *Or.* from concave posterior surface of cricoid cart.; *Ins.* base of arytenoid cart. *Use*, to separate the chordæ vocales and open the rima glottis.

Crico-arytenoideus lateralis, *Or.* side of cricoid cart. near its upper margin; *Ins.* base of arytenoid cart. *Use*, to open the rima glottidis and relax the vocal chords.

Arytenoidei. Fill the concave post. surfaces of the arytenoid cartilages, they consist of oblique and transverse fibres passing from one to the other. *Use*, to approximate the arytenoid cart., close the rima, and relax the vocal chords.

Some muscular fibres are occasionally met with in the thyro epiglottic and aryteno-epiglottic folds, and are named after them. They are of little importance.

The mucous membrane of the larynx forms a portion of the bronchial membrane, and resembles it in most particulars; it is but loosely connected to the subjacent parts by loose areolar tissue.

The *upper opening* of the larynx or *glottis* communicates with the isthmus faucium; triangular, it is bounded in front by the epiglottis, which forms its base, its apex posteriorly at the convergence of the arytenoid cartilages, its sides are formed by the aryteno-epiglottic folds.

The *lower opening* or *rima glottidis*, a narrow aperture, slightly triangular, is bounded on each side by the true chordæ vocales, the base is behind at the arytenoid cartilages, the apex in front, at the retreating angle of the thyroid cartilage, where the vocal chords arise. The rima glottidis is liable to much and constant change; the glottis varies but'little.

The arteries of the larynx are derived from the superior and inferior thyroid; the nerves are branches of the par vagum or pneumogastric nerve; of these the superior laryngeal enters the thyro hyoid membrane, and is distributed to the arytenoid, thyro arytenoid, crico arytenoid lateralis, and crico thyroid muscles, the inferior sends filaments to the same musc. to the crico arytenoid posticus and anastomoses with the superior.

Section V.

The Pharynx.

The pharynx is a musculo-membranous tube, of a funnel shape, the larger extremity being continuous with the mouth and nares, the smaller with the œsophagus; it extends from the base of the cranium to the 4th or 5th cervical vertebra; its length, subject to continual variation, is about four inches and a half; its upper ext. is about two inches wide, and its lower one eleven lines. The following chain of fixed points affords attachment to its muscles. The cricoid and thyroid cartilages, os hyoides, stylohyoid ligament, base of the tongue, infer. maxilla, pterygo-maxillary lig. int. pterygoid plate and petrous bone; behind, they are attached to a strong aponeurosis, which descends from the basilar process and Eustachian tubes.

Constrictor infer.; Or. 1, a triangular space on the cricoid cartilage, between the crico thyroid and arytænoid m.; 2, the oblique line, post. and sup. edge, and smaller cornu of the thyroid cartilage; *Ins.* with its fellow into the raphé on the back part of the pharynx... The *infer.* laryngeal nerve passes beneath its lower edge; the *super.* under its upper edge.

Constrictor medius. Triangular by its apex from the body and cornua of the os hyoides and stylo hyoid lig.; *Ins.* by its base with its fellow into the raphé. its super. fibres are continued by means of a pointed tendon to the basilar process. the stylo pharyngeus m. passes under its upper edge.

Constrictor super, Or. 1, base of the tongue, where it is continuous with the genio hyo glossus, 2, post. extremity of the mylo hyoid ridge; 3, pterygo maxillary ligament; 4, lower third of the inner pterygoid plate, and by an aponeurosis from the petrous bone. *Ins.* tendinous raphé and basilar process. an oval sinus (Morgagni) is left on each side, between its upper fibres and the cranium. The pharynx is also strengthened by fibres from the petrous, basilar, hamular, and styloid process. If the pharynx be opened by an incision along the raphé, there will be seen the velum, isthmus faucium, base of the tongue, epiglottis and seven openings—viz., two of the nares, two Eustachian tubes, the mouth, the glottis, and the œsophageal entrance.

The pharynx is lined literally by mucous membrane, continuous with that of the mouth, larynx, &c. This is covered by a prolongation of the cuticle termed the epithelium, which extends as far as the cardiac extremity of the œsophagus, where it terminates in a fringed margin.

The arteries supplying the pharynx are derived from the pharyngea ascendens branch of the ext. carotid, the tonsillitic and nasal arteries; it is supplied with nerves from the pharyngeal plexus formed by the pharyngeal branch of the par vagum, the glosso pharyngeal and sympathetic nerves.

In deglutition the pharynx is first opened to receive the morsel of food, its cavity is expanded from before backwards by the elevation of the larynx, and from side to side by the stylo pharyngei and palato pharyngei muscle; the morsel of food having descended into it, the constrictor muscles contract from above downwards, and force the aliment into the œsophagus, and thence it passes into the stomach.

The *Velum* is a kind of septum, projecting backwards and downwards from the posterior edge of the palate into the pharynx, between the apertures of the mouth and nares; its muscles, ten in no., are inserted into a dense aponeurosis, usually called the expansion of the tensor palati m., continued from the fibrous portion of the palate, Eustachian tubes, and septum nasi. *The velum is raised to a horizontal line by*

Levator palati, Or. narrow, tendinous from the petrous bone near its summit, and the adjoining portion of the Eustachian tube; *Ins.* broad into the post. edge of the aponeurosis, and with its fellow into the whole length of the median line beneath the motores uvulæ.

Circumflexus palati, Or. 1, scaphoid fossa at the base of the int. pterygoid plate; 2, adjoining portions of the sphenoid bone and Eustachian tube: it first descends, and is then reflected round the hamular process, running horizontally inwards to be lost in the aponeurosis palati ∴ Separated from the latter m. by the sup. constrictor.

Motores uvulæ, two narrow fasciculi, descending along the median line, from the nasal spine to the end of the uvula ∴ behind the levator palat. *The velum is depressed by*

Palato glossus, Constrict, isthma. fauc., ant. pillar of the fauces narrow in the middle, its super. fibres are expanded with those of the following muscle into the inf. surface of the velum, and its inferior are lost with the stylo glossus in the side of the tongue.

Palato pharyngeus. post. pillar of the fauces; *Or.* broad, inf. surface of the palate, and whole length of the median line of the velum; *Ins.* whole length of the post. edge of the thyroid cartilage, and into the pharynx. ∴ The two pillars on each side are separated by the tonsil.

The velum is lined on both surfaces by mucous membrane, and contains numerous glands, vessels, and nerves. Its arteries are derived from the palatine branches of the labial and int. maxillary: its nerves are branches from Meckel's ganglion, &c.

From the centre of its free margin depends the uvula, composed of a few muscular fibres, some glands and cellular tissue; on each side of this the velum is curved or concave, and laterally terminates in two projections formed by the palato glossus and palato pharyngeus muscles; these form the half arches of the palate, and separating inferiorly leave an interval in which is lodged the amygdala or tonsil.

The use of the velum is to prevent the food from passing upwards into the nares during deglutition; in this act it is raised and spread out so as to present an inclined plane forwards. It is also useful in forming the sounds of the voice.

The *amygdala*, or *tonsil*, is lodged in the recess between the half arches of the palate; it is almond shaped, the smaller extremity turned downwards; its structure is a number of mucous follicles, connected by cellular tissue, and covered by mucous membrane.

The *Posterior nares* lie immediately above the velum, each is quadrilateral in shape, and conveys the air from the anterior nares into the isthmus faucium and lungs.

The *opening of the Eustachian tubes* are situated one on each side of the posterior nares, opposite and a little above the post. extremity of the inf. spongy bone. They conduct the external air into the cavity of the tympanum.

CHAPTER III.

THE UPPER EXTREMITIES

Are attached to the trunk by means of the clavicle and numerous muscles; some of them are situated on the ant. and some on the post. surface.

SECTION I.

Muscles of the Anterior and Lateral Parts of the Trunk.
Thorax.

Pectoralis major, Or. 1, sternal half of the clavicle; 2, ant. surface of the sternum; 3, the cartilages of all the true ribs, except the first and last. *Ins.* into the ant. lip of the bicipital groove by a double tendon; the ant. portion of which receives the upper and middle fibres, the posterior the infer. fibres. *Use,* to draw the arm inwards to the side, and rotate it inwardly; the clavicular portion raises the arm; if this be fixed the sternal and costal portions will assist in inspiration.

Pectoralis minor, Or. third, fourth, and fifth ribs, just beyond their cartilages; *Ins.* into the point of the coracoid process ∴ crosses the middle of the axillary vessels and nerves. *Use,* to depress the glenoid cavity inwards, also to assist in inspiration.

Subclavius, fusiform, *Or.* by a flat tendon from the cartilage of the first rib. *Ins.* ext. half of the inf. surface of the clavicle; covered by a strong aponeurosis (costo-coracoid ligament) which arches over the vessels and muscle from its attachment to the clavicle to the coracoid process. *Use,* to depress the clavicle and shoulder inwards, to raise the first rib.

Serratus magnus, Or. by nine or ten digitations from the ten sup. ribs. The first portion arises by one digitation from the first and second rib, is *Ins.* into the sup. angle of the scapula; a second portion by three digit. from the second, third, and fourth ribs, *Ins.* into the base of the scapula; third portion by six digit. from fifth, six, seventh, eighth, ninth, and tenth ribs, *Ins.* into the inf. angle of the scapula ∴ the five inf. digitations intersect with five of the obliquus descend. m. *Ins.* between the subscapularis and the rhomboidei and levat. scapulæ. *Use*, to draw forwards the scapula, to assist in inspiration—the sup. fibres depress the glenoid cavity—the inf. fibres elevate it by rotating the scapula.

Intercostales, eleven internal and eleven external, on each side. They are attached to the opposite edges of each pair of ribs. The fibres of the ext. extend from the transverse processes of the vertebræ to the costal end of the ribs, and by an aponeurosis from that point to the sternum: those of the *internal* extend from the sternum to the angles of the ribs, and from that point by an aponeurosis to the spine. ∴ the intercostal vessels and nerves separate the two layers. ∴ the fibres of the ext. run downwards and outwards in the direction of the obliq. extern. abdom. ∴ the internal in a contrary direction like the ob. int. abd. *Use*, to raise the ribs in inspiration.

Levatores costarum, triangular, twelve on each side, *Or.* pointed from a transverse process above. *Ins.* broad ext. surface of the rib next below. *Use*, to raise the ribs.

Triangular sterni, (serrat. min. antic.) *Or.* post. surface and edges of the lower part of the sternum, xiphoid, and adjoining costal cartilages. *Ins.* by distinct digitations into the inf. margins of the sixth, fifth, fourth, and third costal cartilages. *Use*, to depress the costal cartilages in expiration.

Section II.

Muscles on the Posterior part of the Trunk.

Dorsal muscles—First layer, 2.

Trapezius. Or. 1, thin and aponeurotic from the inner third of the sup. occip. ridge; 2, the twelve dorsal spinous processes; 3, from the last cervical spine, and the *ligamentum nuchæ*, which extends from that process to the occipital protuberance. *Ins.* the spine of the scapula, the acromion process and the ext. third of the clavicle.·.from the same extent of the latter processes arises the deltoid. The tendinous fibres of the two muscles, from the sixth cervical to the third dorsal vertebræ, represent an ellipsis. *Use,* to elevate the shoulder, to depress the head backwards and to one side; the sup. fibres raise the glenoid cavity directly, the inf. fibres rotate the scapula and so raise the glenoid cavity; to fix the scapula and thus antagonize the serratus magnus in inspiration.

Latissimus dorsi, Or. from the six inferior dorsal spinous processes, by a strong fascia, which is confounded with that of the int., obliq. and transversalis m.; from the spines of the lumbar vertebræ and sacrum, and the post. third of the crista ilii: lastly, by so many digitations from the three or four last ribs; *Ins.* the inner and post. edge of the bicipital groove in front of the teres major.·.its costal attachments interdigitate with the digitations of the ext. obliq.; sometimes an additional slip arises from the inf. angle of the scapula. *Use,* to draw the arm backwards and inwards, so as to rotate the palm of the hand inwardly; to compress the side of the thorax in respiration; to assist in raising the ribs if the arm be previously fixed.

Second Layer, 3.

Rhomboidei, 1, *R. major, Or.* four or five sup. dorsal spines; *Ins.* base of the scapula, from its spine to its inf. angle; 2, *R. minor, Or.* last cerv. spine; *Ins.* base of the scapula, between the two roots of the spinous process. *Use,* to draw backwards the scapula, and depress the glenoid cavity.

Levator anguli scapula, Or. post. transverse tubercles of the three or four sup. cerv. vert.; *Ins.* into the angle of the scapula, and its post. edge as far as the spine. Between the splenius colli and the scaleni and rect. capit. ant. *Use,* to raise the sup. angle of scapula, and depress the glenoid cavity.

Third Layer, 4.

Serratus post. superior, Or. lig. nuchæ, first cervical, and second or third upper dorsal spines; *Ins.* by slips into the second, third and fourth ribs, beyond their angles. *Use,* to raise these ribs.

Serratus post. inferior, Or. two last dorsal and three upper lumbar spines; *Ins.* the four last ribs, just beyond their angles. A fine but strong aponeurosis is stretched from one muscle to the other; which internally is attached to the spinal processes, externally to the ribs, beyond their angles. *Use,* to depress the lower ribs in expiration.

Splenius colli et capitis, both muscles *arise* from the two or three upper dorsal and the last cervical spines, and from the ligament. nuchæ, as far as the third cervical spine; *Ins.* the *s. colli* into the post. trans. tubercles of the first and second, sometimes the third cervical vertebræ; the *s. capitis* into the post. edge of the mastoid process, and adjoining third of the rough surface between the two occipital curved lines. ∴ S. colli, at its insertion, is behind the L. ang. scap. *Use,* to bend the head and neck backwards.

Fourth Layer.

The fourth layer is entirely covered by the above muscles and their fascia.

Longissimus dorsi. The mass of muscle bearing this name fills up the interval between the spinous processes and ribs, where it is bound down by the serratus fascia; in the lumbar region it is ensheathed by the superficial and middle lumbar fascia; in this situation the muscle constitutes the common origin of three distinct muscles and their cervical prolongations. It is attached, fleshy, to the whole surface of the sacroiliac groove, and to the anterior surface of a dense fascia, which is attached to the post. third of the crista ilii and its posterior spinous processes, and to those of the sacrum, lumbar, and three last dorsal vertebræ.

Sacro lumbalis (the external division.) *Or.* from the common mass; *Ins.* by distinct tendons into the six last ribs, near their angles; here it would cease, but the *musculi accessorii*, arising by distinct tendons from all the ribs successively from the last, join the muscle so as to enable it to deposit similar tendons on the angles of the six upper ribs, and to ascend in the neck under the name of

Cervicalis ascendens. It is formed by the fourth or fifth upper *musculi accessorii*; *Ins.* post. transverse tubercles of the fourth, fifth, and sixth cervical vertebræ.

Longissimus dorsi (middle division,) *Or.* from the common mass; *Ins.* internally into the *transverse* processes of all the dorsal, and into the *spinous* processes of the five or six upper dorsal vertebræ; and externally into all the ribs between their tubercles and angles. It is prolonged into the neck as the

Tranversalis cervicis. Its *origin* is formed by *mus. access.* from the third, fourth, fifth and sixth dorsal transverse process; *Ins.* into the post. tubercles of the third or fourth upper cervical transverse processes, between the cerv. ascend. and trachelo mastoideus.

Spinalis dorsi (inner division,) *Or.* from the innermost portion of the common mass, which reaches as high as the ninth dorsal vert.; *Ins.* into the spinous processes of the nine superior ribs; its cervical prolongation is the

Spinalis colli. It *arises* by *mus. accessorii* from the transverse processes of the five dorsal vertebræ, to be *Ins.* into the spinous processes of the six last cervical.∴ it is covered by the complexus. A deeper seated portion of the spinalis dorsi, from the diminished length of its fasciculi, is called

Semi spinalis dorsi. It *arises* from the dorsal transverse processes, from the fifth to the eleventh; *Ins.* to the spinous processes of the three super. dorsal and two inf. cervical vertebræ. Deeper still, the fasciculi are much shorter, and are called

Multifidus spinæ; *Or.* by slips from all the transverse pr., between the first sacral and the third dorsal vert.; each slip is *Ins.* into the first, second, or third spinous process next above; the last is attached to the spine of the axis.

Complexus, *Or.* 1, five or six upper dorsal transverse processes; 2, four last cervical articulating processes; 3, spinous processes of the last cervical, and two upper dorsal. *Ins.* inner half of the rough surface between the occipital ridges.

Trachelo mastoideus (complexus minor,) *Or.* trans. processes of the four last cervical and four upper dorsal vertebræ; *Ins.* back part of the mastoid process, beneath the splenius.

Fifth Layer.

Inter spinales. Double, in the cervical; single, in the lumbar and dorsal vertebræ. Indistinct in the latter.

Inter transversalis. Double, in the cervical; single, in the lumbar and dorsal vertebræ. Indistinct in the latter. The principal *use* of the deep spinal muscles is to fix the spinal column, and move its several portions slightly; they are antagonistic to most of the other muscles attached to the spine.

Rectus posticus major, Or. narrow from the spine of the axis; *Ins.* broad into the inf. occipital transverse ridge. *Use*, to depress the head posteriorly.

Rectus posticus minor, Or. narrow from the spine of the atlas; *Ins.* broad into the occipital bone, behind the foramen magnum. *Use*, to depress the head backwards.

Obliquus inf., Or. narrow from the spine of the axis; *Ins.* narrow, end of the transverse process of the atlas. *Use*, to rotate the atlas and head on the second vertebræ.

Obliquus super., Or. narrow, from the transverse process of the atlas; *Ins.* broad into the rough surface between the occipital ridges just behind the mastoid process. *Use*, to bend the head backwards and to one side.

Section III.
Superior Extremity.

Immediately beneath the skin and subjacent cellular tissue the muscles of the arms and side of the thorax will be found, covered with a tolerably dense layer of fascia, which descends from a spine of the scapula, the acromion process, and clavicle, to invest them. It is best marked on the inner side of the arm, where it covers the brachial vessels and nerves, and on the ant. surface of the biceps musc. It is firmly attached to the lower margins of the pectoralis major and latissimus dorsi, and passes from these across the floor of the axilla, so as to close and protect its cavity. Inferiorly, the brachial aponeurosis is attached to the intermuscular septa, and the condyles of the humerus.

Muscles of the Shoulder, 5.

Deltoideus, Or. 1, from the whole extent of the post. border of the spine of the scapula; 2, the acromion process; 3, the ext. third of the clavicle. *Ins.* narrow, into a triangular surface on the outer side of the hu-

4*

merus a little above its middle, between the two heads of the brachialis anticus m. ; separated from the pectoralis major by the descending branch of the thoracic acromialis art. and the cephalic vein. *Use*, to raise the arm from the side, or to depress the shoulder. The ant. fibres draw it forwards; the post. backwards.

Supra spinatus, Or. the inner two-thirds of the fossa supra spinat.; *Ins.* super. facette of the larger tubercle of the humerus. *Use*, to raise and abduct the arm.

Infra spinatus, Or. inf. surface of the spine, and the dorsum of the scapula as far as the post. of the two ridges on its inf. costa; *Ins.* middle facette of the larger tubercle of the humerus. *Use*, to raise the arm, and draw it backwards.

Teres minor Or. inf. costa of the scapula, between its two ridges; *Ins.* inferior facette of the larger tubercle of the humerus. The tendons of these three muscles cover the head of the humerus, and are intimately adherent to the capsule of the joint. *Use*, similar to the preceding. The principal use of the three capsular muscles is to support the articulation, and keep the head of the humerus in situ.

Teres major, Or. form a flat quadrilateral surface on the inf. angle of the scapula; *Ins.* post. margin of the bicipital groove, behind the tendon of the latissimus dorsi. *Use*, to draw the arm downwards, backwards, and to the side; to rotate it inwards; and when the arm is fixed, to draw forwards the inf. angle of the scapula, and elevate the glenoid cavity.

The two teres muscles, in proceeding to their insertions, leave an opening between them, which is divided into two by the passage of the long head of the triceps; the anter. one transmits the post. circumflex artery and nerve; the other the r. dorsalis scapulæ of the infra scap. art. and nerve.

Subscapularis Or. subcapular fossa and its margins; *Ins.* the inner or smaller tubercle of the humerus to keep the head of the humerus and glenoid cavity in close apposition.

Muscles of the Arm: in front, 3; Flexors.

Coraco brachialis, Or. from the tip of the coracoid process in conjunction with the short head of the biceps, to which it is attached for some way down; *Ins.* into the inner side of the humerus, opposite to the insertion of the deltoid; it first covers and then runs along the outer side of the art. brachial: perforated by the ext. cutan. nerve. *Use,* to raise, draw forwards, and rotate the arm outwards.

Biceps, Or. the short head, from the tip of the coracoid process with the coraco brachialis m.; the long head, from the upper edge of the glenoid cavity where its tendon bifurcates, its division descending on the margins of that cavity to meet similar divisions on the long head of the triceps. *Ins.* by a strong flat tendon, which is twisted half around, into the tubercle of the radius: just before the tendon sinks to its insertion, it gives off the *semilunar fascia,* which crosses the pronator teres, to strengthen the fascia of the fore arm. ·. The long head is enclosed in the capsular ligament of the shoulder joint, but outside of the synovial membrane, which accompanies it for some way down the groove. ·. its inner edge constitutes an unerring guide to the brachial artery. *Use,* to flex the fore arm on the arm, to make tense the fascia, and rotate the radius outwards; also to depress the glenoid cavity and scapula.

Brachialis anticus, Or. 1, by two angular slips which embrace the insertion of the deltoid; 2, from the condyloid ridges of the humerus, and anterior surface between them, down to the elbow joint. *Ins.* coronoid process of the ulna, and rough surface beyond. *Use,* to flex the fore arm.

Back part of the Arm, 1, Extensor.

Triceps, Or. 1, its long head, by a flat tendon from the lower edge of the neck of scapula, where it bifurcates to join the divisions of the biceps; 2, middle head, from the post. and ext. surface of the humerus, between the spiral groove and insertion of the teres minor, and from the ext. condyloid ridge and aponeurotic septum, as far as the condyle; 3, the small head, from the int. and post. surface of the humerus, *below* the spiral groove and insertion of the teres major m., and from the int. condyloid ridge and intermuscular septum. *Ins.* by a strong aponeurosis into the olecranon.·.the spiral groove commences above the small head, winds in front of the long head, and below the super. attachments on the middle head. *Use*, to extend the fore arm, and depress the neck of the scapula. The long head separates the two spaces beneath the inf. margin of the scapula; its second and third heads are separated by the musculo spinal or radial nerve.

In the dissection of the muscles of the shoulders and arm, numerous vessels and nerves are met with. The arteries are branches of the subclavian, the superior scapular, and posterior scapular; of the axillary, the thoracica acromialis, thoracica suprema, glandulares thoracica longa, or submammary, the subscapular, and the anterior and posterior circumflex; of the brachial, the superior profunda, inferior profunda, and anastomatica magna.

The superior, posterior, and subscapular arteries, as their names imply, surround the margins of the scapula, and sending branches to the muscles on both of its surfaces, anastomose freely together, and thus establish an intimate connexion between the subclavian and axillary arteries; and by no means of their subanastomoses with the carotid arteries above and the brachial arteries beneath, are of essential importance in tying any one of these vessels. The remaining branches of the axillary, excepting the circumflex are found connected with the pectoral muscles, and cavity of the axilla; these anastomose freely also with the vessels about the shoulder externally, and the thoracic and intercostal arteries internally, as to keep up the circulation when one or other of the large trunks is tied. The circumflex arteries surround the neck of the humerus, and are chiefly lost in the deltoid.

The *nerves* of this region are principally derived from the brachial plexus; they are anterior and posterior thoracic to the thoracic muscles, the circumflex to the deltoid, the suprascapular to the fossa of the same name, the subscapular to the subcap. muscle, and sending a few filaments to the dorsal surface of the scapula, and lastly, the nerves to the arm and forearm, viz., on the outer side the ext. cutaneous and median, on the inner the int. cutaneous and ulnar, and posteriorly the musculo-spiral or radial. Some small nervous filaments are met with in the subcutaneous cellular tissue; these are chiefly on the inner side, and consists of branches from the internal cutaneous, and the nerves of Wrisberg, derived from the 2nd and 3rd intercostals.

The principal *veins* are the cephalic and basilic, and the brachial trunk.

The *Axilla*, or *Axillary* cavity is triangular, its apex superiorly at the coracoid process; the base inferiorly formed by the axillary fascia, stretching from the pect. major to the latissimus dorsi; it is bounded in front by the pectoral muscles, behind by the subscap. musc., internally by the serratus magnus, externally by the neck of the humerus; it contains the axillary artery and vein, and their principal branches, and the axillary plexus of nerves, the continuation of the brachial plexus embeded in cellular tissue and a number of lymphatic glands.

In the female subject, the *Mammary Glands* are found one to each side resting on the pectoralis major muscle to which they are loosely connected by areolar tissue. The mammary gland is hemispherical in shape, convex in front, flat posteriorly: it is composed of a number of distinct *lobules*, separated by condensed cellular tissue, and some adipose substance. Each lobule is composed of a number of *granules*, from which the lacteal ducts proceed, and gradually uniting open upon the free extremity of the nipple. This is a slightly conical projection, composed of the integuments, lacteal ducts, and erectile areolar tissue; it is surrounded by an *areole* of the integuments, generally of a brownish colour, but in the young female, unimpregnated, is of a pinkish hue; on this open a number of sebaceous follicles.

The mammary gland is supplied with blood from the thoracic branches of the axillary, the intercostal, and the internal mammary. The nerves are derived from the brachial plexus and the intercostals. Its absorbents are very numerous, and communicate with the axillary, cervical, and thoracic glands.

Section IV.

Of the fore-arm.

The fore-arm extends from the elbow to the wrist-joint; it is covered by the integuments, beneath which lies a quantity of loose cellular tissue in which are found in front, the superficial veins—viz., the cephalic on the outer, the basilic on the inner side, and median in the centre, and superficial nerves branches of the int. and ext. cutaneous; on the back part, also, some superficial veins and nerves, the latter branches of the ext. cutaneous and musculo-spiral nerves.

Beneath this is the *ant. brachial aponeurosis*, or *fascia*, exceedingly dense, investing the muscles, and sending septa between so as to separate them from each other; it is attached superiorly to the condyles of the humerus, intermuscular septa and tendon of the biceps, on the inner side to the spine of the ulna, and inferiorly to the ant. and post. annular ligaments of the wrist.

MUSCLES OF FORE-ARM.   47

Muscles of the Fore Arm, in front, 8.

Superficial set, 5, *which arise from the inner condyle.*

Pronator teres, Or. 1, the inner condyle and intermuscular septa; 2, from the coronoid process of the ulna. *Ins.* outer edge of the radius, at its middle. The median nerve passes between its two origins. *Use,* to flex the fore arm, and rotate the radius inwards.

Flexor carpi radialis, Or. inner condyle and septa; *Ins.* base of the second metacarpal bone. *Use,* to flex the fore arm and hand.

Palmaris longus, Or. inner condyle and septa; *Ins.* annular ligament and palmar fascia. *Use,* to make tense the palmar aponeurosis and flex the hand.

Flexor carpi ulnaris, Or. 1, inner condyle; 2, inner edge of the olecranon; 3, by an aponeurosis from the upper two-thirds of the inner edge of the ulna. *Ins.* pisiform bone, its fibres strengthening the ligament which unites that bone to the os pyramidale. the ulnar nerve passes between the first and second origins: its inner edge is a guide to the lower two-thirds of the ulnar artery. *Use,* to flex the hand.

Flexor sublimis perforatus, Or. 1, inner condyle; 2, the inner edge of the coronoid process of the ulna; 3, upper part of the oblique ridge of the radius. *Ins.* into the fingers by four tendons, which split at the end of the first phalanx, to give passage to the tendons of the deep flexor; each division then winds round the edges of the corresponding tendon of the latter muscle, so as to form a kind of sheath for it, and both are then inserted close together into the fore part of the second phalanx. *Use,* to flex the fingers on the hand, and this on the wrist.

Deep seated Muscles, 3.

Flexor profundus perforans, Or. 1. ant. three-fourths of the fore part of the ulna and inner edge of the coronoid process; 2, inner half of the interosseous ligament; 3, by a slip from the radius, just below its tubercle. *Ins.* by four tendons, each of which perforates a corresponding tendon of the last m., and is inserted into the last phalanx. *Use*, to flex the fingers on the hand, and this on the wrist.

Flexo longus pollicis Or. 1, ant. surface of the radius, from its tubercle to the upper edge of the pronator quadratus; 2, outer half of the interosseous ligament. *Ins.* last phalanx of the thumb. *Use*, to flex the thumb.

Pronator quadratus Or. lower fifth of the anterior surface, and inner edge of the ulna; *Ins.* ant. surface and inner and outer edges of the lower fourth of the radius. *Use*, to pronate the radius and hand.

Muscles on the outer edge, 4.

Supinator radii longus, Or. ext. condyloid ridge of the humerus and intermuscular septum, from the spiral groove to within two inches of the condyle; *Ins.* into the styloid process of the radius. ∴ the lower two thirds of the radial artery run along the inner edge of its tendon. *Use*, to supinate the hand.

Extensor carp. radialis longior, Or. from the lower part of the external condyloid ridge, and the external condyle; *Ins.* post. surface of the carpal end of the second metacarpal bone.

Ext. carp. rad. brevior, Or. post. part of the external condyle, with the common extensor: *Ins.* post. surface of the carpal end of the third metacarpal bone.

Supinator radii brevis, Or. 1, from the outer condyle by the common tendon; 2, from the ext. lateral and annular ligaments: 3, from a ridge on the ulna and triangular fossa in front of it. *Ins.* into the fore-part, and outer and back part of the radius, in a line leading obliquely from above its tubercle to the ins. of the pronator teres. Pierced by the posterior branch of the spiral nerve: the biceps tendon passes through the fibres of the muscle where they cover the tubercle of the radius. The *use* of these muscles is indicated by their names.

Muscles on the back of the Fore-Arm.

Superficial set, 4.

Extensor digitorum communis, Or. 1, by a tendon, which it shares with the ext. carp. r. b., ext. carp. uln., and ext. min. digit., from the ext condyle; 2, from the intermuscular aponeuroses. *Ins.* into the phalanges by four tendons, each of which first sends off an expansion from its edges to cover the dorsum of the first phalanx, and then, dividing into three slips sends one to the upper end of the second phalanx, and one on each side of the first joint to the upper end of the last phalanx; the three ext. tendons are united by transverse slips, before they reach the phalanges.

Extensor minimi digiti, Or. from the outer condyle by the common tendon; *Ins.* back part of the phalanges of the little finger, like the corresponding tendon of the common extensor.

Extensor carpi ulnaris, Or. 1, outer condyle by the common tendon; 2, post. surface and outer edge of the ulna; *Ins.* back part of the carpal end of the fifth metacarpal bone. The uses of the preceding muscles are indicated by their names.

Anconæus, Or. by a distinct tendon from the outer condyle; *Ins.* 1, into the outside of the olecranon where it is continuous with the triceps; 2, triangular surface on the upper fourth of the ulna. *Use,* to extend the fore-arm.

Deep-set, 4.

Extensor metacarpi pollicis, Or. from the ulna, interosseous ligament, and radius, just below the supin. brevis muscle; *Ins.* back part of the carpal end of the first metacarpal bone.

Extensor primii nternodii, *Or.* 1, radius and interosseous ligament below the last m.; 2, slightly from the ulna. *Ins.* back part of the carpal end on the first phalanx.

Extensor secundi internodii, *Or.* broad from the middle third of the back of the ulna, and slightly from the interosseous ligament. *Ins.* back of the carpal end of the second phalanx. The use of the preceding muscles is indicated by their names.

Indicator, *Or.* back part of the ulna and interosseous ligament below the last muscle; *Ins.* by slips into the phalanges of the index finger, like the common ext. *Use*, to extend the forefinger, as in pointing.

In the dissection of the muscles of the fore-arm, the vessels and nerves of this region are exposed. The arteries are the *radial, ulnar,* and *interosseous*.

The *radial* artery proceeds from the bifurcation of the brachial art. opposite the coronoid process of the ulna, along the outer and ant. surface of the fore-arm, at first deep but inferiorly superficial, and then winds around the ext. lat. lig. of the wrist to pass between the first and second metacarpal bones into the palm of the hand. Its principal branches are the radial recurrent, and superficialis volæ; the rest are muscular branches.

The *ulnar* art. lies on the inner and ant. surface of the fore-arm, and like the preceding, becomes more superficial near the wrist; it here passes above the annular lig. and enters the palm of the hand. Its principal branches are the ant. and post. ulnar, recurrent and interosseous; the rest are chiefly muscular.

The *interosseous* art. is a branch of the ulnar; it sinks deep, lies on the interosseous membrane, and terminates in the pronator quadratus muscle. Its principal branch is the post. interosseous, which passes to the back of the fore-arm between the ant. oblique lig. and the interosseous membrane.

The nerves met with in the fore-arm are the musculo-spiral, or radial, ulnar and median.

The *musculo-spiral*, or *radial* nerve, joins the radial at the junction of the upper and middle thirds of the fore-arm, runs along its outer side, and at the junction of the middle and lower thirds winds around the radius beneath the supinator longus tendon to the back of the wrist and hand; a little beneath the elbow it sends off its largest branch, which perforates the supinator brevis to supply the muscles on the back of the fore-arm.

The *ulnar* nerve accompanies the ulnar art. on its inner side, and enters with it the palm of the hand; its principal branch proceeds from it some distance above the wrist, and bends round the ulna to the back of the wrist and hand.

The *median* nerve accompanies the brachial art., sinking into the fossa beneath the elbow, passes between the two heads of the pronator quadratus, descends to the wrist, where it lies behind the palmaris longus tendon, and enters the palm of the hand, beneath the annular lig.

The deep-seated veins are the venæ comites, two of which accompany each art., lying on either side; they unite generally at the elbow to form the brachial vein, although the basilic frequently joins the rest above the joint.

Section V.

Of the Hand.

Beneath the integuments of the palm of the hand and a quantity of adipose tissue, lies the *palmar aponeurosis*, or *fascia*. This is exceedingly dense; triangular in shape, it proceeds posteriorly from the ant. annular lig. forwards and divides into four fasciculi, each of which subdivides into two, to allow of the passage of the flexor digitorum tendons, the digital vessels and nerves, and pass to be inserted into the lateral ligaments of the metacarpo-phalangeal articulations.

Beneath this lie the muscles, vessels, and nerves, with the exception of the palmar brevis, which lies superficial to it.

Muscles of the hand, 19.

Those on the outer edge, 4.

Abductor pollicis, Or. from the scaphoid bone and annular ligament of the carpus; *Ins.* with the outer head of the short flexor.

Opponens pollicis, Or. from the trapezium and annular ligament: *Ins.* whole length of the outer edge of the metacarpal bone.

Flexor brevis pollicis, *Or.* by two portions, 1, from the trapezium, lower edge of the annular ligament, and sheath on the flex. carp, rad. m.; 2, from the os magnum. *Ins.* by two heads, each enclosing a sesamoid bone, into both sides of the first phalanx, with abductor on one side, and the abductor m. on the other.

Abductor pollicis, Or. the whole length of the ant. surface of the third metacarpal bone; *Ins.* with the inner head of the short flexor. The use of these muscles is indicated by their name.

Muscles on the inner edge, 4.

Palmaris brevis, quadrilateral *Or.* annular ligament and inner edge of the palmar fascia; *Ins.*, its short parallel fibres are lost in the skin over the inner edge of the hand. *Use*, to corrugate the integ., on the inner side of the hand.

Abductor minimi digiti, *Or.* pisiform bone; *Ins.* inner edge of the base of the first phalanx.

Flexor brevis, d. minimi, thin when it exists, *Or.* lower edge of the carpal annular lig. and unciform process is soon identified with the former. The ulnar art. and nerve separate them at their origins.

Opponens dig. minimi, Or. the same as the short flexor; *Ins.* whole length of the inner edge of the fifth metacarpal bone. The use of these muscles is indicated by their names.

Muscles in the middle of the Hand, 11.

Lumbricales, Or. fore and outer parts of each tendon of the flexor profundus; *Ins.* with the corresponding interossei into the edge of the dorsal expansion on the same side. *Use*, to assist the flexor tendons, or to act with them.

The interossei. If the median line or axis of the third metacarpal bone be taken as a fixed point, the palmar interossei may be considered as adductors, and the dorsal as abductors, both sets arise from the opposite edges of two metacarpal bones and are inserted into the edges of the dorsal expansion. The *dorsal interossei*, four in no., are attached, the first, or *indicator*, to the radial edge of the index finger, the second, to the radial edge of the medius, the third to the ulnar edge of the medius, and the fourth, to the ulnar edge of the ring finger. The *palmar interossei*, three in no., are attached, the first, to the ulnar edge of the index, the second, to the radial edge of the ring finger, and the third, to the radial edge of the little finger.

The relations of the parts about the wrist-joint are as follows. A. *Back of the joint*, 1, styloid process of the radius and ins. of the supinator longus, covered by the tendons of ext. metacarpi et primi internodii pollicis; 2, the tendon of the ext. secund. internodii. In their triangular interval between this m. and the two latter, may be felt the radial artery just before it passes between the metacarpal bones, and the tendons of the two radial; extensors; 3, the tendon of the exten. indicis, those of the common extensor, and that of the exten. min. digiti; 4, the extensor carpi ulnaris, and the styloid process of the ulna. B. *Fore part of the joint*, commencing at the same point, 1, surface of the radius, supporting the radial artery; 2, tendon of the flexor carpi radialis, and immediately beneath it that of the flexor proprius pollicis; 3, tendon of the palmaris longus, the tendon of the flexor profundus being felt in the interval; 4, flexor carpi ulnaris and pisiform bone, the tendons of the flexor sublimis supporting the ulnar artery and nerve in the interval. The median nerve passes under the annular ligament, imbedded in the flexor tendons, and at the ulnar edge of the flexor c. radialis.

The arteries and nerves met within the palm of the hand are the superficial and deep palmar arches.

The *superficial palmar arch of arteries* lies immediately beneath the palmar aponeurosis; it is formed by the ulnar art. and superficialis volæ branch of the radial, and supplies three and a half fingers, counting from the little finger.

The *superficial palmar arch of nerves* accompanies this, lying beneath it; this is formed by the median nerve, and supplies three and a half fingers, counting from the thumb, the remaining finger and a half being supplied by the ulnar nerve.

The *deep palmar arch of arteries and nerves* lies beneath the flexor tendons on the interossei muscles; the first is formed by radial art. with the deep branch of the ulnar; it supplies one finger and a half counting from the thumb; the latter is formed by the deep branch of the ulnar, and is lost chiefly in the interossei muscles.

CHAPTER IV.

THE TRUNK

is composed of two large cavities, the thorax and abdomen, each of which contains important viscera, most of these belonging to the organic life of the individual. The pelvis assists in forming the trunk inferiorly. It will be described with the lower extremities.

SECTION 1.

The Thorax

is a large conical cavity giving lodgment to and protecting the great central organs of respiration and circulation, namely, the *lungs, heart, aorta, pulmonary artery, vena cava, thoractic duct,* &c. It is formed by the twelve dorsal or thoracic vertebræ posteriorly; the ribs posteriorly, laterally, and in front; the sternum and costal cartilages anteriorly; its apex is truncated, and forms an oval aperture superiorly, bounded by the first dorsal vertebra posteriorly, the first rib laterally, and the upper end of the sternum and first costal cartilage in front; through this pass *upwards* the sterno hyoid. sterno thyroid, and longus colli muscles; the arteria innominata, left subclavian, and carotid arteries; cone of the pleura, right and left recurrent nerves, and thoracic duct; *downwards,* the venæ innominatæ, internal mammary, and superior intercostal arteries; pneumogastric, phrenic, recurrent and sympathetic nerves, trachea, and œsophagus.

The *base* of the thorax is formed inferiorly by the diaphragm; through this pass the aorta, inferior vena cava, and vena azygos, pneumogastric, phrenic, splanchnic, and sympathetic nerves, œsophagus and thoracic duct. The *perpendicular diameter* is greatest posteriorly and laterally, least anteriorly; the *trans-*

verse is greatest inferiorly, least superiorly; the *antero posterior* is greatest inferiorly from the lower end of the sternum backwards to the spine.

The THORACIC CAVITY is divided into five compartments by the reflexion of the pleura, of these two are on the mesial lines, the anterior and posterior mediastinum; one, on the left side, the pericardium, and two laterally the pleura cavities for the lungs.

The *pleura* is a serous membrane which lines most of the interior of the thorax, one on each side, each being distinct from the other, so as to form two separate membranes. The right is the larger but shorter.

Each pleura passes upwards into the neck as high as the fourth cervical vertebra, forming its cone; inferiorly it lines the upper surface of the diaphragm, and here forms the ligamentum latum pulmonis; intermediately it is reflected in the following manner; from about the junction of the sternum and costal cartilages it passes outward, lining the inner surface of the ribs, until it reaches to the side of the dorsal vertebra; it here turns forward, forming the side of the posterior mediastinum, and passes outwards on the post. surface of the root of the lungs, lines these organs, passing in between their lobes, round the ant. surface of the root of the lungs, then lines the lateral surface of the pericardium, from which it approaches its fellow of the opposite side, passes forwards and outwards, forms the lateral boundary of the anterior mediastinum, and arrives at the junction of the sternum and costal cartilages. The pleura is divided as it lines the lungs or ribs into the pleura pulmonalis and costalis; the latter is the stronger of the two.

The *anterior mediastinum* inclines downwards and a little to the left side; it extends from the upper to the lower extremity of the sternum, triangular in shape, its base turned forwards is formed by the post. surface of the sternum and a little by the costal cartilages, its apex turned backwards is formed by the convergence of the pleuræ, which bound it laterally; it is slightly contracted in the centre. Its *contents* are the thymus gland, the triangularis sterni muscles, and the int. mammary arteries.

The *thymus gland* is largest in the fœtus, when it nearly fills the ant. mediastinum; it consists of two lobes, divided by an oblique fissure. Its *use* is wholly unknown. In the adult it degenerates into cellular tissue.

The *lungs* are the great organs of respiration. Each lung is conical in shape, and corresponds to the pleural cavity; the apex fits into the cone of the pleura; its base, concave, rests on the upper surface of the diaphragm; its anterior surface, concave, is applied against the pericardium; the outer and posterior surfaces are convex. The right lung is the larger but shorter, and consists of three lobes; the left has but two lobes. These lobes are separated by the great fissure which runs obliquely downwards and forwards, so as to divide each lung into an interior smaller, and a post. larger lobe; from the middle of the fissure proceeds a small fissure on the right side, cutting off the third from the superior lobe.

The *root of the lung* is formed by an assemblage of the several structures entering or leaving the organ— viz., the bronichial tube, pulmonary art., and pulmonary veins, surrounded by cellular tissue, and covered by the pleura. The order of parts in each lung is from behind forwards, bronchus, artery, veins; from above downwards, on the right side, bronchus, art., veins: on the left, art., bronchus, veins; these are enveloped by the pulmonic plexus of nerves, best marked on the posterior surface, and formed by the pneumogastric and sympathetic nerves. A number of dark-coloured glands, *bronchial glands*, are found around the root of each lung.

The *structure* of the lungs is composed of bronchial tubes and air cells, pulmonary arteries and veins united by fine areolar tissue,

The *bronchial tubes* are the continuation of the trachea, from the bifurcation of which, opposite the second dorsal vertebra, they descend obliquely downwards and outwards to the root of each lung. The right bronchus is less oblique than the left; it passes behind the superior cava to the root of the lung; the left bronchus passes through the arch of the aorta, in front of the thoracic aorta, to the root of its lung. The bronchial tubes divide into numerous branches, which ultimately terminate in forming the *air cells*, which form a slight expansion resembling a cluster of grapes. A bunch or cluster of the air cells form the lobules, and these united form the lobes. These lobules are united by fine areolar tissue, the *interlobular cellular tissue*, in which the pulmonary arteries and veins run to their destination.

The *structure* of the bronchial tubes is the same as that of the trachea, except in their small branches and air cells, which want the cartilaginous structure. The structures of these consist of an internal lining mucous membrane, an ext. cellular coat, and a thin layer of muscular fibres between: some deny the existence of these.

The *pulmonary arteries* enter the lungs at their root, and divide the right into three, the left into two branches, to supply the lobes; they gradually break down into capillary vessels, which ramify on the outer surface of the air cells, and so conduct the black blood to be oxygenated by the atmospheric air.

The *pulmonary veins* arise from the capillaries of the pulm. art., and unite on each side into two large vessels, which cross inwards, and open into the posterior surface of the left auricle of the heart; and thus conduct the art. blood into this organ. The right are longer and larger than the left.

The *pulmonic plexus* sends its numerous filaments along the posterior surface of the bronchial tubes to their minutest ramifications, and finally terminates in the air cells.

THE TRACHEA.

The *absorbents* of the lungs are numerous, and terminate in the bronchial glands.

The *bronchial arteries* are small vessels which arise from the thoracic aorta to supply the lungs with nutrition. Their blood is returned by the bronchial veins. There is a free anastomosis between them and the pulmonary vessels.

The lungs in the healthy adult are of a mottled gray colour, are light and spongy; in the young subject, after breathing, they are of a reddish colour, before breathing are dark red; in the diseased state they are frequently nearly black, from a deposit of carbon in their structure.

The lungs are light and spongy, and float in water.

The *trachea* descends from the termination of the larynx, opposite about the fourth cervical vertebra, downwards and a little to the right side of the mesial line of the neck, enters the cavity of the thorax between the ant. and post. mediastinum, and divides opposite the second dorsal vertebra into the right and left bronchial tubes.

In the neck, the trachea is covered in front by the sterno hyoid and thyroid muscles, and partly by the thyroid gland; it rests posteriorly on the œsophagus, which inclines to its left side; on each side of it is the carotid sheath and contents, the recurrent and sympathetic nerves. In the thorax it is crossed by the thymus gland, and left vena innominata, and supports the upper part of the arch of the aorta. The phrenic, pneumogastric, and sympathetic nerves descend to its outer side, and behind it is the œsophagus and spinal column. The right vena innominata descends on its right side, the thoracic duct ascends on its left side, but post. to it.

The *structures* of the trachea consist of cartilaginous rings, fibrous tissue, mucous membrane, and muscular fibres. The tracheal rings form but three parts of a circle, being deficient posteriorly. They are from eighteen to twenty in number, and vary in size; the first is the largest; they are connected by fibrous membrane. The trachea is completed posteriorly, where the rings are deficient, by three layers of tissue, the ext. is fibrous, some consider elastic; the next is transverse muscular fibre: and the internal is the lining mucous membrane; between these are some small glands which open on the mucous membrane.

The pericardium and heart occupy the middle mediastinum, and will be found under the head of the Vascular System.

The *posterior mediastinum* extends nearly on the mesial line in front of the spine, from the second or third to the tenth dorsal vertebra, triangular in shape, its apex, turned forwards, is formed by the convergence of the pleuræ, its base backwards by the spinal column; its contents are, the œsophagus and thoracic aorta to the left side, the vena azygos to the right, the thoracic duct in the centre, the splanchnic nerves are found in its lower part.

The *œsophagus* descends from the pharynx to the left side behind the trachea, through the arch of the aorta, crosses a little in front of the thoracic aorta, perforates the diaphragm, and terminates in the cardiac extremity of the stomach: on its surfaces are found, a little above the diaphragm, the pneumogastric nerves, the left being anterior. The œsophagus is composed of longitudinal muscular fibres, lined by mucous membrane; some circular fibres are described at its cardiac extremity, but their existence in the human subject may well be questioned.

The *thoracic aorta* lies to the left side of the spine, but inferiorly comes forward in front, and enters the abdomen through the aortic opening.

The *vena azygos* is formed by the junction of the lumbar veins in front of the second lumbar vertebra; it ascends through the aortic opening of the diaphragm, passes upwards in the posterior mediast. to its right side, and finally hooking round the root of the right lung, terminates in the sup. vena cava: in this course it receives the eight or nine inferior intercostal veins, and generally the left vena azygos, which crosses from the left side behind the aorta to join it; this is formed by the six inf. intercostal veins of the left side; the remaining intercostal veins form the superior intercostal veins, and terminate in the right vena innominata; the left receives the superior six intercostal veins on its side, and terminates in the left vena innominata.

The *thoracic duct* is the trunk of the absorbent system; it is formed in front of the third lumbar vertebra by the junctions of the lymphatics from the lower extremities and pelvis, and passes upwards in front of the spine; it escapes from the abdomen through the aortic opening of the diaphragm, between the aorta to its left and the vena azygos to its right side, enters the post. mediastinum, ascends on the mesial line of the spine, inclines to its left side, passes behind the arch of the aorta, and ascends into the neck, behind the left carotid and int. jugular vein, as high as the sixth cervical vertebra; it now turns downwards and inwards, and terminates in the left subclavian vein near its junction with the int. jugular.

In its course the thoracic duct occasionally divides into two or more branches, which unite again into one trunk. Its structure consists of two coats, an external fibrous and an internal serous, which is thrown into numerous folds forming valves.

There is generally a small thoracic duct on the right side, which conveys the absorbents from the right side of the head and neck and right arm into the right subclavian vein.

The *splanchnic nerves* are formed from the last six dorsal ganglia of the sympathetic; they perforate the diaphragm, and terminate in the semilunar ganglion and renal plexus.

Section II.

The Abdomen,

the largest cavity in the human body, is bounded above by the diaphragm, below by the pelvis, behind by the lumbar vertebræ and quadratus lumborum muscles, laterally and in front by the abdominal muscles. Unlike the thorax, its parietes are chiefly muscular, and thus exert a necessary pressure on the abdominal viscera.

The abdomen is divided, for the purposes of description, into three principal regions, namely, the upper or *epigastric*, the central or *umbilical*, and the inferior or *hypogastric*. A line drawn from the inferior edge of the costal cartilages across to the opposite side separates the two superior, and a similar transverse line drawn between the ant. sup. spinous processes of the illium separates the two inferior; each of these regions is further subdivided into three by the lineæ semilunares, which extend from the cartilages of the eighth or ninth rib downwards and inwards towards the pubis.

The *epigastric* thus consists of a central, or epigastric region, and a right and left hypochondriac; the *umbilical* of a central umbilical and a right and left lumbar; and the *hypogastric* of a central hypogastric and a right and left iliac region. The lower portion of this region is further subdivided into a central pubic and a right and left inguinal region.

Anterior and Lateral Muscles of the Abdomen, 5.

Obliquus externus, *Or.* the eighth or ninth infer. ribs by fleshy digitations, of which five intersect with the serratus magnus, and three or four with the latissimus dorsi; *Ins.*, 1, fleshy into the ant. two-thirds of the crista ilii; 2, the rest of the muscle ends at a line from the ant. sup. spinous p. to the ensiform cartilage in an aponeurosis, by which it is *ins.* into the whole length of the linea alba, the lower portion of this aponeurosis (*Poupart ligament*) extends from the ant. sup. spin. p. across the femoral vessels to be *ins.* by one slip (*int. pillar of the ext. ring*) into the symphysis pubis, by another (*ext. pillar of the ext. ring*) into the spine of the pubis; 3, by a broader portion (*Gimbernat's ligament,*) which is folded under the cord, presenting a sharp lunated edge towards the femoral vessels, into the spine of the pubis, and an inch along the pectineal line .·. the reflected portion of the aponeurosis forms the ant. and inf. walls of the inguinal canal .·. a space exists between its post. edge and the ant. edge of the latis. dorsi m. which has been the seat of hernial protrusion, (Petit.) *Use*, to depress the ribs and compress the abdominal viscera, to raise the pelvis if the thorax be fixed, or to depress them forwards if the pelvis be fixed.

Obliquus internus, Or. 1, from the fascia lumborum; 2, ant. two-thirds of the crista ilii; 3, from the ext. two-thirds of the grooved or internal surface of Poupart's ligament. *Ins.* the upper portion, fleshy into the edges of the four inf. costal cartilages; the middle terminates at the linea semilunaris in the middle abdominal aponeurosis, by which it is continued to the linea alba; the portion from Poupart's ligament (with the lower fibres of the transversalis) arches over, and then behind, the cord (Cooper,) to be *ins.* into the body of the pubis, behind the ext. ring, and into the pectineal line .·. the last portion, with the transversalis, forms the upper and posterior walls of the ing. canal. (Fasciæ.) *Use*, nearly similar to the preceding; it rotates the trunk in the opposite direction, and thus co-operates with the ext. obliquus of the other side of the body.

Cremaster, Attachments, 1, ext. two-thirds of Poupart's ligament; 2, behind that ligament at its pubic attachment; the intervening fibres descend in successive loops to be *ins.* into the cord, and to surround the tunica vaginalis, so as to sling the testicle ∴ the testicle, in its passage beneath the lower edge of the obliquus int. m., draws down with it the muscular fibres to form the cremaster (Cloquet;) the cord passes between the lower fibres of the transversalis, from which also the muscle is derived. (Guthrie.) *Use*, to suspend the testis.

Transversalis, *Or.* 1, ext. two-thirds of Poupart's ligament; 2, ant. three-fourths of the crista ilii; 3, fascia lumborum; 4, inner surfaces of the six last ribs, by digitations, which intersect with the diaphragm.— *Ins.* the upper part terminates at the linea semilunaris in the *post.* aponeurosis, by which it is continued to the linea alba: that from Poupart's ligament passes over and behind the chord to be *ins.* with the int. obliq. into the pubic end of Poupart's ligament and ilio pectineal line. Another portion passes behind and then under the cord, in front of the fascia transversalis, to Poupart's ligament; thus forming the poster. and part of the inf. wall of the inguinal canal. (Cooper.) *Use*, to assist the obliq. muscles in respiration and the compression of abdm. viscera.

Rectus Abdominis, *Or.* by a double tendon from the pubis, between its spine and symphysis; is *ins.* by three portions: 1, ensiform cartilage and costo xiphoid lig.; 2, fifth costal cartilage; 3, sixth and seventh costal cartilages ∴ The muscular fibres are partially interrupted by three or four zigzag tendinous intersections which adhere to the forepart of the sheath. *Use*, to approximate the pelvis and thorax.

Pyramidalis, triangular, sometimes absent, *Or.* by its base, from the pubis and ant. pubic ligament. *Ins.* by its apex into the linea alba, sometimes as high as half-way towards the umbilicus. *Use*, to assist the preceding.

These muscles are enclosed in a fibrous sheath, formed by the tendons of the obliquus and transversalis muscles. This sheath extends from the xiphoid and costal cartilages superiorly, to the crest of the pubis, into which it is implanted; it is formed by the ext. oblique tendon, and one layer of the int. oblique passing in front of the rectus musc. and behind by the second layer of the int. obliquus and transversalis, except from half-way between the umbilicus and pubis to this bone, where the whole of the tendons pass in front of the rectus; this muscle is intersected by three or four irregular transverse lines: the lineæ transversæ one at the umbilicus, one at the xiphoid cartilage, one between these, and one sometimes beneath the umbilicus. By this intersection each portion of the muscle may act independently of the rest. In the sheath behind the muscle the epigastric art. ascends to anastomose with the int. mammary.

The *lineæ semilunares* are the oblique lines on the outer edge of the rectus, formed by the junction of the oblique and transversalis tendons. The linea alba is the tendinous white line which extends from the xiphoid cartilage superiorly to the pubes inferiorly; it is formed by the intermingling of the fibres of the oblique and transversalis muscles, is thinner and wider above than below the umbilicus, where it is particularly dense.

Post. Abdominal Muscle.

Quadratus Lumborum, *Or.* ileo lumbar lig. and two inches of the crista ilii; *Ins.* 1, lower edge of the inner half of the last rib; 2, by four processes into the four upper lumbar transverse processes .·. enclosed in a sheath like the rectus; vide, *Fasciæ*. *Use*, to draw down the ribs.

Superior Abdominal Muscle.

Diaphragm. A moveable septum between the thoracic and abdominal cavities, composed of a central tendinous portion, surrounded by muscular fibres. The tendon is trilobate, and placed nearly in the centre of the muscle; above, it supports the heart being in the *adult* firmly united to its fibrous pericardium; below, it covers the centre of the liver. The middle and largest lobe is behind the xiphoid cartilage; the right, the next in size, and the left, the smallest, are directed backwards and outwards on each side. The post. margin of the tendon is notched, and gives attachment to the fixed portion of the muscle; the anterior edge, from which the anterior or moving portion, arises, is convex and concentric with the infr. margin of the thorax; *Or. posterior or fixed portion;* 1, *the crura.* The left crus arises narrow tendinous from the bodies and intervertebral subs. of the three upper lumbar vertebræ; the right, which is more in front, arises from the corresponding vertebræ, and the intervertebral substance next beyond. Opposite the first lumbar vert. they become fleshy, and divide into two portions, which are separated by the splanchnic nerves; the outer portions, are *ins.* into the post margin of the tendon: the inner decussate in front of the vert. col. to form the aortic opening, diverge to form the œsophageal opening, and then pass on to the central tendon. 2, Another portion arises from the false and true ligam. arcuata on each side to be *inserted* into the remainder of the notch. The *ant. portion, Or.* from the whole of the convex edge of the tendon *Ins.* base of the xiphoid cartilage, and by six or seven digitations into the inner surfaces and upper edges of the six last ribs, where it intersects with the m. transversalis.

The edge of the aortic opening is tendinous; it transmits the aorta, vena azygos, and thoracic duct; the œsophageal opening is muscular; it transmits the œsophagus and the vagus nerves. The opening for the vena cava is tendinous quadrangular, and tubular, and is placed at the junction of the right and middle lobe of the tendon. ∴ The cellular tissue of the mediastinum, and that between the ant. abdominal muscles, communicating through an opening on each side of the xiphoid cart.

Section III.
The Abdominal Viscera.

The abdomen contains the organs of digestion viz.: the stomach and intestines, the liver, pancreas, and spleen, and in addition, the kidneys, supra-renal capsules, and ureters; the abdominal aorta, inf. vena cava, vena porta, thoracic duct, and numerous lymphatic vessels and glands, the sympathetic nerve, and several smaller vessels and nerves.

The abdomen is lined internally by the *Peritoneum*; this is the largest serous membrane in the body, and lines the inner surface of the abdominal parietes as well as most of the contained organs.

The *reflections* of the peritoneum may be traced as follows—from the umbilicus it passes upwards, lining the inner surface of the tranversalis and recti muscles, to the lower margin of the thorax; it here bends backwards, lining the inf. concave surface of the diaphragm, sinking deep into the hypochondriac region; on the left it is reflected on the back part of the splenic vessels, to the post. surface spleen, it winds round and invests this organ, then passes on the ant. surface of its vessels; in the centre it is reflected on the stomach so as to enclose this organ, and pass from its margin downwards to form the gastro colic omentum; on the right side it is reflected from the umbilicus, enclosing the umbilical vein, and forming the suspensory ligament of the liver; it invests this organ, and passing from its inferior surface encloses the hepatic vessels, and forms the gastro-hepatic fold or omentum; from the diaphragm the peritoneum descends on the liver, and forms the lateral ligaments of this organ; from the convex or inferior border of the stomach the two layers of peritoneum unite to form the gastro-colic omentum, which passes downwards in front of the transverse arch of the colon, descends lower on the left than on the right side, becomes reflected on itself, so as to consist of four layers; the two posterior layers ascend until they reach the transverse colon; they here separate so as to enclose this intestine, and from its concave margin unite and descend to form the transverse meso colon; opposite the duodenum this fold separates into an ascending and descending layer, enclosing the inferior portion of this intestine; the ascending layer passes in front of the lower and middle portions of the duodenum and of the pancreas to the posterior surface of the right lobe of the liver, where it is continuous with the peritoneum lining this organ and the posterior layer of the lower or gastro-hepatic omentum; the *descending* layer passes downwards into the lumbar region, where it becomes continuous with the right and left meso colon; in the centre the inferior layer of the transverse meso colon passes in front of the spinal column and the large blood-vessels resting upon it, and is thence reflected downwards and forwards so as to enclose the small intestines and the mesenteric vessels and form the mesentery.

Inferiorly the peritoneum descends from the umbilicus in the centre, enclosing the urachus, and forming a slight projection, it then descends, lining the posterior surface of the rectus muscle to near the pubes; it than passes backwards, sinks into the pelvis, lines the upper posterior and part of the lateral surface of the bladder as low down as near to the posterior extremity of the vesiculæ seminales and prostate gland, whence it is reflected on to the anterior surface of the rectum and the sides of the pelvis forming the posterior and lateral false ligaments of the bladder; the peritoneum now ascends, investing the middle and upper thirds of the rectum, forming the meso rectum, whence it passes on each side to line the iliac fossæ, and becomes continuous with that lining he abdominal parietes. In the female the peritoneum is reflected from the posterior surface of the bladder on the upper and back part of the vagina, whence it passes upwards on the uterus so as to invest this, and form on each side the broad ligaments of this organ, and from it passes upon the rectum as in the male subject. On either side the peritoneum covers partially the cæcum and signoid flexure of the colon, and binds them down in the iliac fossa.

On each side of the mesial line inferiorly, the peritoneum is thrown into an oblique fold by the degenerated umbilical artery, and thus are formed the inguinal pouches one on each side of the umb. art.

The peritoneum in the female does not form a shut sac, as an opening exists in it, where it covers the fimbriated extremities of the Fallopian tubes; here the serous and mucous membranes are continuous.

The principal folds formed by the peritoneum are: The *gastro hepatic*, or *lower omentum*, consists of two layers, and extends from the transverse fissure of the liver to the concave margin of the stomach; it contains between its layers the hepatic art. to the left side; the ductus choledochus to the right, and the vena porta behind and between both; these are surrounded by cellular tissue, and form the capsule of Glisson, behind which is the foramen of Winslow leading into the great bag of the omentum.

The *gastro colic*, or *great omentum*, the meso colon, meso rectum, and mesentery have been already sufficiently described. The great bag of the omentum extends from the foramen of Winslow to the bottom of the great omentum; it is bounded in front by the lesser omentum, the stomach, and the descending layer of the great omentum; and posteriorly by its ascending layer.

The hollow viscera of the abdomen consists of the stomach, the small and large intestines.

The STOMACH, the principal organ of digestion, is situated in the left hypochondriac and epigastric regions; it extends from its cardiac or œsophageal extremity downwards and forwards, and near its pyloric extremity turns a little upwards and backwards; it is curved and irregular in shape, and has two orifices, two curves, and two extremities.

The *cardiac orifice* communicates with the œsophagus, and is nearly circular; its *pyloric orifice* is situated towards the right side, anterior to the cardiac; it is also circular, and communicates with the duodenum; the left, or great bulging extremity of the stomach, lies in front of the spleen; the right, much smaller, terminates in the pylorus; the concave, or lesser edge of the stomach is turned upwards and backwards, and has attached to it the lesser omentum, the coronary vessels and nerves run along it. The great, or convex edge, is directed downwards and forwards, and has attached to it the great omentum, the epiploic vessels and nerves run along this border; the anterior surface of the stomach is convex, the posterior nearly flat.

The stomach is composed of four coats, the *serous, muscular, nervous* or *fibrous*, and the *mucous;* the *serous* coat is derived from the peritoneum, and has been already described. The muscular layer consists of three sets of fibres; the longitudinal are derived from the fibres of the œsophagus, and are best marked along the lesser and greater curves; the circular fibres surround this organ at the junction of its right third with its left extremity; they sometimes contract so as to give to the stomach an hour-glass shape; the irregular or oblique fibres are situated chiefly around the great bulging extremity.

The *nervous* or *fibrous* coat of the stomach lies beneath the preceding; it is composed of condensed cellular tissue, and strengthens the stomach considerably; the mucous coat lines the interior of the stomach; it is of a rose pink colour, and is thrown into irregular folds, the rugæ of this organ; at the pylorus this forms a distinct valve, having beneath it well-marked circular muscular fibres.

The stomach is supplied with blood from the coronary arteries, a branch of the cœliac axis, which runs along its lesser curve, the right gastro-epiploic, a branch from the hepatic, and the left gastro-epiploic, a branch of the splenic. Its nerves proceed from the pneumogastric or par vagum and the solar plexus.

The SMALL INTESTINES consist of the duodenum jejunum, and ileum. The *duodenum* is about twelve inches in length, and extends from the pylorus to the root of the mesentry; it lies chiefly in the right hypochondrium, and reaches into the umbilical and left lumbar regions, it forms an irregular horse-shoe curve, in the concavity of which lies the head of the pancreas, and consists of three portions—the superior transverse, the vertical and the inferior transverse—the first passes upwards and to the right side as far as the neck of the gall-bladder; the second descends along the right side of the spinal column in front of the right kidney; the third portion passes transversely across the spine, in front of the second lumbar vertebra, and behind the sup. mesenteric art. terminates in the jejunum.

The duodenum has four coats, the external or serous is only partial, the superior portion is entirely covered by peritoneum, the middle wants this membrane posteriorly, and the inferior portion is uncovered, lying between the folds of the meso colon; its muscular fibres are chiefly circular, its fibrous coat resembles that of the stomach, its mucous coat is thrown into numerous folds at its lower third, forming the *valvulæ conniventes*. The ductus communis choledochus and the pancreatic duct open into the concave margin of the duodenum, near to and a little above the angle of junction between its descending and inferior transverse portions, generally on one of the valvulæ conniventes; its orifice is marked by a slight thickening of the membrane,

The *jejunum* forms two-fifth of the small intestines,

the *ileum* the remaining three-fifths; their structure is nearly similar, consisting of an external serous coat a muscular coat, composed chiefly of circul. fibres, a fibrous coat, and a mucous lining membrane, thrown into numerous folds, the valvulæ conniventes. These are most numerous in the jejunum, least so in the lower part of the ileum.

The small intestines are supplied with blood, the duodenum chiefly from the gastro duodenalis art., the jejunum and ileum from the superior mesenteric art. Their nerves are derived from the solar plexus.

In the interior of the small intestines are numerous glands. They are isolated in the upper part, forming the glandulæ solitaræ, or the glands of Brunner, but in the lower part are collected in oval groups forming the glandulæ agminatæ, or the glands of Peyer.

At its lower extremity the ileum ascends from the pelvic cavity, and terminates opposite the right sacroiliac symphysis in the cæcum or caput coli.

The LARGE INTESTINES consist of the cæcum, colon, and rectum.

The *cæcum*, or *caput coli*, is lodged in the right iliac fossa, is irregular in shape, and larger than any other part of the large intestines; it receives the end of the ileum at its internal and posterior side, some distance above its lower extremity, and is continuous with the colon; its coats consist of a partial serous covering, deficient posteriorly, longitudinal, and a few circular muscular fibres, a fibrous coat, and a lining mucous membrane; its interior is irregular, and presents at its junction with the ileum a double valve, the *ileo cæcal valve;* the inferior portion of this valve is the larger, and somewhat vertical, the superior is more horizontal; the structure of these valves is a fold of mucous membrane enclosing some muscular fibres. The *appendix vermiformis*, a small worm-like intestine, proceeds from the left and posterior surface of the cæcum; its cavity is small, and communicates with the end of the cæcum.

THE RECTUM.

The *colon* extends from the cæcum to the rectum, and is divided into the *right ascending*, the *transverse, left descending*, and the *sigmoid flexure*. The ascending colon is situated in the right lumbar region, in front of the right kidney and renal vessels, the transverse colon passes across the spine to the left side, forming an arch, the convexity forwards, and joins the descending colon; this is situated in the left lumbar region in front of the left kidney and renal vessels, and joins inferiorly the sigmoid flexure; this is lodged in the left iliac fossa resting on the psoas, quadratus lumborum, and iliacus internus muscles. The structure of the colon is similar to that of the cæcum, in all, the longitudinal muscular fibres are collected into three longitudinal bands; these proceed from the appendix vermiformis, and are lost in the rectum; one of these is anterior, the other two are on the posterior surface; on their outer surface are the *appendices epiploicæ*, folds of peritoneum enclosing some adipose tissue. The large intestines are supplied with blood from the inf. mesenteric art.; their nerves are derived from the sympathetic.

The *rectum* descends into the pelvis from the sigmoid flexure of the colon; it proceeds from the left sacro iliac symphysis, at first downwards and inwards towards the mesial line of the sacrum, it then bends downwards and forwards towards the perineum and beneath the bladder, and finally turning a little backwards, it terminates in the anus; concave in front, it receives the posterior surface of the bladder, and rests in the hollow of the sacrum; on each side of it are the internal iliac vessels and sacral plexus. The structure of the rectum consists of a *serous coat;* this is only partial, the upper third is entirely covered, the middle third wants this covering on its posterior surface, and the lower third is wholly uncovered by it; a *muscular coat* composed of reddish longitudinal fibres, which above the anus assume a circular shape, forming the deep sphincter ani musc., a *fibrous coat*, well marked. and lastly a *lining mucous membrane;* this is thrown into longitudinal folds in the undisturbed state. Some have described also transverse folds, but they do not exist.

The *arteries* of the rectum are derived from three sources, the *superior* hemorrhoidal from the inf. mesenteric, the *middle* from the internal iliac, and the *inferior* from the internal pudic; its nerves are derived from the sacral and sympathetic. The absorbents of the intestines are extremely numerous, those of the small intestines absorb the chyle, and are hence called lacteals; they pass through the mesenteric glands contained in the folds of the mesentery, and terminate in the receptaculum chyli, the commencement of the thoracic duct; the absorbents from the large intestines pass through the lumbar glands, and terminate also in the thoracic duct.

The glandular viscera of the abdomen subservient to the function of digestion are the liver, pancreas, and spleen.

The LIVER is situated in the right hypochondriac and epigastric regions, of greater extent transversely than vertically; its anterior surface, concave, is divided into two by the falsiform ligament, and is in contact with the diaphragm; its posterior surface, irregularly concave, rests on the pyloric end of the stomach, the duodenum, and upper extremity of the right kidney; its upper margin is obtuse and connected to the diaphragm by folds of the peritoneum, and the inf. vena cava; its lower margin is thin, and notched near its centre by the umbilical vein: its lateral margins are rounded.

The liver is divided into two lobes, the right and left, by the attachment of the falsiform lig. on its convex surface, and by the longitudinal fissure lodging the umbilical vein, and ductus venosus, on its concave surface; the left lobe is the smaller, and is not subdivided. From the junction of the superior and middle thirds of the longitudinal fissure, the *transverse fissure* proceeds to the right side through the middle third of the liver; it lodges the two branches of the vena porta, the hepatic art., and hepatic duct; the horizontal fissure runs from the anterior to the posterior margin of the liver, and lodges the umbilical vein, and its continuation, the ductus venosus.

Besides these fissures there are two depressions on the concave surface of the liver, one for the vena cava, sometimes a canal, between the lobulus Spigelii and right lobe; another for the gall bladder to the right of the horizontal and in front of the transverse fissure. The right kidney and colon sometimes mark this surface.

On this surface are the three lobules of the liver—viz., *lobulus Spigelii*, pyramidal in shape, its apex free, having the longitudinal fissure to its left, the vena cava to its right, and the transverse fissure in front; a tail-like process, *lobulus caudatus*, connects it with the rest of the right lobe; *lobulus quadratus* is bounded to the left by the horizontal fissure, to the right by the gall bladder, superiorly by the transverse fissure, in front by the thin margin.

The liver has two coats, a serous and a fibrous; the *serous* or *peritoneal* is partial, as it does not cover the organ at the fissures, behind the gall bladder and vena cava, and within the coronary ligament; it forms the right and left lateral, coronary, and falsiform ligaments; the first two connects the respective lobes to the diaphragm, the second surround the orifices of the vena cava hepat., the last has been already described. The *fibrous* coat is tolerably well marked, and passes into the interior of the liver with its vessels. The structure of the liver consists of a number of granules, of a brownish yellow colour, united by cellular tissue.

Mr. Kiernan has described the structure of the liver as composed of lobular bodies, which are found in all parts of the organ, but in the centre are of an angular shape, while towards the circumference they acquire a more rounded form; running down their middle is the *intralobular* vein, into which still more minute branches pour their blood; at the base of the lobules these veins open into a larger one, called *sublobular;* several of these join together, and by their union form the hepatic vein, which terminates in the vena cava; around the intralobular vein ramify the vena porta, ductus hepaticus, and hepatic art., external to which is the capsular covering; in passing to and from these lobules the vessels ramify between them, and form *interlobular* vessels; the portal vein in the vaginal sheath gives off numerous branches, which, previous to entering between the lobules, form the *vaginal plexus*.

The liver is supplied with blood by the hepatic art. and vena porta, which pass transversely across the organ, each of them having a coating of cellular tissue derived from the capsule of Glisson, and hence collapse when cut across; the blood is returned by the venæ cavæ hepaticæ, which unite into three or four large trunks, and terminate in the vena cava where it passes through the diaphragm; these vessels have no cellular covering, remain open when cut across, and take a longitudinal course.

The *bile* is conducted from the liver by the biliary ducts; these arise from the granules, and finally unite so as to form two ducts, one for each lobe, which again unite and form the hepatic duct; this escapes from the transverse fissure and unites with the *cystic duct* to form the *ductus communis choledochus*, about three inches in length; it conveys the bile into the duodenum.

The *gall bladder* is lodged in a sulcus on the under surface of the liver, to the right of the lobulus caudatus; pyriform in shape, its larger ext. is free, and projects beyond the thin margin of the liver, its smaller ext. terminates in the cystic duct; this, about an inch and a half in length, unites with the hepatic duct to form the ductus communis choledochus. The structure of the gall bladder is composed of three coats; the ext. serous is a partial coat and binds it down *in situ;* the second is fibrous, and the third is mucous; this is honeycomb in appearance, and is tinged with the bile. The structure of all these ducts is the same, namely fibro-mucous, as are all the excretory ducts of the body except the thoracic duct.

The PANCREAS lies across the spine behind the stomach and between the two layers of the transverse meso-colon; it is composed of a larger extremity or head, which lies in the concavity of the duodenum; a lesser ext. or tail, reaching to the spleen, and a central portion, or body; its structure resembles that of the salivary glands.

The *Pancreatic duct* is seen on its posterior surface; it arises by numerous radicles, which unite and form the trunk; this conveys the pancreatic fluid into the duodenum. Its structure is fibro-mucous. The pancreas is supplied with blood from the splenic and the pancreatico duodenalis art.

The SPLEEN is situated in the left hypochondrium, behind the great bulging ext. of the stomach, and above the left kidney. It is a soft spongy mass, of a crescentic shape; the concavity, turned inwards, receives the vessels and nerves. The convexity is in contact with the abdominal parietes beneath, and the diaphragm above. The coats of the spleen are a serous and fibrous, the former nearly perfect, the latter weak and thin; its internal structure is composed of shreds of cellular tissue filled with blood. The spleen is supplied with blood from the splenic art.; its nerves are derived from the sympathetic.

The principal vessels of the abdomen are the aorta and its branches, the vena cava inferior, and the vena porta. (See Vascular System.) The principal nerves are the symphathetic and the lumbar plexus, lying imbeded in the psoas mag. musc. (See Nervous System.)

Section IV.

Hernia.

The anatomy of hernia is intimately connected with that of the abdominal parietes, and may well be considered here.

Inguinal hernia is so called from its appearance in the inguinal region or groin; of it we have two varieties, one named *oblique* inguinal hernia, from the oblique direction which the tumour takes between the abdominal muscles, following the course of the spermatic cord, whence it is also occasionally called spermatocele—the other is called *direct* inguinal hernia, from its protruding directly forwards through the external abdominal ring.

Femoral hernia appears in the upper inner and anterior part of the thigh, where it descends beneath Poupart's ligament on the inner side of the femoral vessels.

Umbilical hernia protrudes through the umbilicus, taking the course of the umbilical cord.

The term *Ventral* hernia indicates the protrusion of one or more of the viscera through some other region of the abdomen than those specifically mentioned; the situations where this species of hernia most frequently occurs, are at the linea alba, the lineæ semilunares and transversæ. The other forms of the disease are of minor importance.

Oblique Inguinal Hernia.

In this form of the disease, the hernia takes the course of the spermatic canal with the spermatic cord; that is, it escapes from the abdomen at the internal abdominal ring, about midway between the spine of the ilium and the symphysis pubis, descends obliquely forwards and inwards, escapes throughout the external abdominal ring, and finally descends into the scrotum.

On raising the skin a dense layer of cellular substance is brought into view—this is the *superficial fascia* of inguinal hernia. This is perfectly continuous with the common subcutaneous cellular tissue beneath the integuments in other regions of the body, and differs but little from it, except in being a little denser in structure, and of more importance from its connexion with hernia. *Above*, the superficial fascia is continuous with the cellular tissue covering the lower surface of the thorax, and upper part of the abdominal muscles; *inferiorly* it descends over Poupart's ligament into the thigh, and there becomes continuous with the superficial fascia of the femoral region; nearer the pubes it passes over the spermatic cord in the male, and descends into the scrotum and perineum, where it identifies itself with the loose cellular tissue of these regions. On the *mesial* line, the superficial fascia descends from the linea alba, on the dorsum of the penis, and here forms a tolerably dense structure, the false suspensory ligament of the penis, which connects this organ to the abdominal parietes. In the female it is loaded with a quantity of adipose tissue and descends into the labium. The cutaneous surface of the superficial fascia is rough and cellular, and is intimately connected to the integuments; its deep surface is more compact and smooth, and is applied against Scarpa's fascia, which thus separates it from the abdominal muscles. The use of this fascia is to allow of the free motions of the integuments on the abdominal muscles, and to assist these in the support of the abdominal viscera.

Some anatomists have divided the superficial fascia into several layers; this is quite an arbitary division. and without any practical advantage, as the number of layers varies in almost every subject, and depends much on the dexterity of the operator; its thickness also varies much in different individuals, in some being but a few in others many lines in depth. In it we meet with several superficial arteries, veins, nerves, and lymphatic glands. The *arteries* are the external epigastric, and branches from the external circumflexa ilii, and pubic arteries. They arise from the femoral artery a little below Poupart's ligament, ascend in front of the ligament, and are lost in the superficial fascia and integuments. The external epigastric is the largest, it ascends obliquely inwards, towards the umbilicus, where it terminates. The *nerves* are small, and are derived from branches of the lumbar plexus. The *lymphatic glands* are four or five in number; they take a nearly transverse course, parallel to Poupart's ligament, some lying a little above, others somewhat beneath the ligament. They are named the *superior* inguinal glands, to distinguish them from another set of inguinal glands, met with below Poupart's ligament more numerous, and arranged in the vertical direction; they are contained in a sheath or capsule formed in the superficial fascia. The superior set communicates with the absorbents of the genital organs. The *inferior* inguinal glands, on the contrary, communicate with the absorbents of the lower extremity.

On raising superficial fascia a distinct layer of cellular tissue "Scarpa's fascia," is exposed. This is firmly attached to, and appears to arise from the fascia lata, about half an inch below Poupart's ligament, whence it ascends over the ligament, to be gradually lost on the outer surface of the external oblique muscle: it is sometimes described as the deep layer of the superficial fascia, with which it is so intimately connected, as to give to this the appearance of being firmly attached to Poupart's ligament; they are, indeed, with difficulty separated.

On Scarpa's fascia femoral hernia rests when it turns up over Poupart's ligament. These fasciæ, excepting the skin, form the most superficial coverings of an inguinal hernia.

On raising Scarpa's fascia the outer surface of the external oblique muscle is brought into view: this is here perfectly tendinous, in order that it may the more effectually support the abdominal viscera; it forms, with the fellow of the opposite side, a broad sheet of tendon, extending directly across the lower part of the abdomen. Each tendon is somewhat triangular in shape: one margin is turned inwards, and unites with its fellow on the mesial line, to form the linea alba; a second is directed upwards and outwards, slightly concave, and gives attachment to the fleshy fibres of the muscle; the third margin is turned downwards and outwards and here presents a thickened condensed cord, named Poupart's ligament.

The tendinous fibres of the external oblique muscle are distinct and well marked: by far the greater number take a direction downwards and inwards; a few may be observed to decussate with these taking an oblique course upwards and inwards. As we trace them towards the pubes, they increase in strength and density, and having arrived within about an inch and a half of this bone, they separate into two bands, and thus leave an interval between them, named the *external abdominal ring*. The bands are named the pillars of the ring. One of them, the *internal*, lies *anterior* and *superior* to the other; it is a broad flat sheet of tendon, which, passing inwards, descends in front of the symphysis pubis, into which it is implanted, decussating with its fellow of the opposite side. A few fibres pass from it to be continuous with the fascia lata of the opposite side; others may be traced descending on the dorsum of the penis, contributing to form the true suspensory ligament of this organ. The *external* pillar of the ring, is stronger and shorter than the other, is a continuation of Poupart's ligament: it is a round fibrous cord; it descends inwards, and is inserted into the spine or tubercle of the pubes; on it the spermatic cord rests as it escapes through the ring. Here the external pillar is excavated on its superior surface (as will be seen in a future stage of the dissection) to accommodate the cord, and thus preserve it from injury.

Inferiorly, or towards the thigh, the *descending* fibres of the external oblique muscle become condensed and form *Poupart's ligament*, a dense fibrous cord, which descends inwards from the anterior superior spinous process of the ilium to the pubes, into which it is implanted by two attachments; one, already described as the external pillar of the ring, is inserted into the spine or tubercle of the pubes; the other is an expansion from this, and forms a broad thin ligament, which ascends obliquely inwards, to be inserted into the internal extremity of the linea innominata; it is named *Gimbernaut's ligament*, or third insertion of the crural arch. It lies superior and posterior to the external pillar of the ring, or second insertion of Poupart's ligament. It is more immediately connected with femoral hernia, with which we shall hereafter describe it more minutely.

Near the ilium Poupart's ligament is thin and weak, but it increases much in density as it aproaches the pubes. Its *inferior* margin slightly concave, and directed somewhat forwards, gives attachment to the iliac portion of the fascia lata. To its *superior* edge are connected the external and internal oblique, and transversalis muscles, and the transversalis and iliac fasciæ.

A few tendinous fibres of the external oblique muscle decussate with those just described; they are the *intercolumnal bands*. They arise from near Poupart's ligaments, ascend inwards, forming curves convex towards the pubes, and are gradually lost in the tendon as they approach the linea alba; towards the ilium they are indistinct, but near the pubes they become strong and well marked; in general one stronger than the rest rounds off the apex of the external abdominal ring, and thus assists in restraining the increase of this opening. The use of the intercolumnal bands is to prevent the separation of the descending fibres of the external oblique, between which, intervals occasionally exist, exposing the fibres of the internal oblique muscle.

Hitherto the external abdominal ring has not been sufficiently apparent; it is concealed from view by a layer of cellular tissue, the external spout-like or Camper's fascia, or intercolumnal fascia, which descends from the edges or margins of the ring, on the spermatic cord, upon which it is insensibly lost. Some describe this fascia as being derived from the intercolumnal bands. Scarpa considers it to be a prolongation of the fascia which has received his name. It is so intimately connected with all these structures, that it is impossible to assign it to any distinct origin. It forms the third covering of an inguinal hernia, whether oblique or direct.

On dissecting off this fascia, the *external abdominal ring* is exposed. This may be now seen to be, as already described, an opening in the external oblique tendon, formed by the divergence of its *descending* fibres; it transmits the spermatic cord in the male, the ligamentum teres in the female. It is triangular in shape; the, *apex*, directed upwards and outwards, and rounded off by the intercolumnal bands; its *base* inferiorly is formed by the crest of the pubes; its *internal* margin, the longest, is formed by the internal pillar of the ring, its *external* margin by the external pillar; its greatest length corresponds to that of its internal margin, and varies from one inch to one inch and a half; it is the outer or anterior opening of the inguinal or spermatic canal; through it an inguinal hernia emerges, to descend into the scrotum.

If the spermatic cord be now cut across, a little beneath the ring, and gently raised, it will be seen that it has escaped through this aperture, resting not on the bone, but on the external pillar of the ring, and that, behind the external abdominal ring, some fibrous structures exist which assist in preventing the protrusion of hernia through this opening, directly from behind, or, in other words, prevent the formation of a direct inguinal hernia; these are Colles's fascia, and the conjoined tendons.

Colles's fascia, or the *triangular ligament of inguinal hernia*, is liable to much variety, being in some subjects well marked, in others altogether wanting; it lies behind the external abdominal ring, in front of the conjoined tendons, and partially concealed by the internal pillar of the ring. It is formed by a few fibres, which are derived from the posterior surface of the external oblique tendon of the opposite side; the ligaments of either side, therefore, decussate. Its base is implanted into the crest of the pubes; one margin is towards the linea alba, the other is free, and looks upwards and outwards.

The *conjoined tendons* lie behind, and to the outside of the preceding: they are the united tendons of the internal oblique and transversalis muscles, which here pass in front of the rectus muscle; to be inserted into the upper part of the symphysis and crest of the pubes; the external margin of the conjoined tendons is prolonged beyond the outer edge of the rectus muscle; it here is inferiorly implanted into the linea innominata, where it is continuous with Gimbernaut's ligament. Above this the transversalis fascia is implanted into it; here, indeed, the two structures frequently appear to be contiouous one with the other.

From the great strength and transverse extent of the conjoined tendons, and their position immediately behind the external abdominal ring, they contribute much to the security of the abdomen in this situation, and thus prevent the formation of direct inguinal hernia.

The *inguinal or spermatic canal* is an oblique canal or passage between the abdominal muscles, which transmits the spermatic cord in the male, the ligamentum teres in the female. It commences at the *internal abdominal ring*, as yet concealed from view by the lower margin of the internal oblique muscle. This is an opening in the transversalis fascia, situated about midway between the spine of the ilium and the symphysis pubis, but somewhat nearer the former, about three-quarters of an inch above Poupart's ligament; from this the canal descends forwards and inwards between the oblique muscles, and terminates at the external abdominal ring. Its length is about one inch and a half, if we measure the distance between the two nearest points of the two rings; three inches, if we measure between the two furthest points of these openings, and two and a half inches, if we measure between the intermediate points.

The inguinal canal is bounded, *above*, by the lower margins of the internal oblique and transversalis muscles, *below* by Poupart's ligament, *in front* by the external oblique tendon and the lower margin of the internal oblique, *behind* by the transversalis fascia, the conjoined tendons, Colles' ligament, and occasionally a few fibres of the internal oblique.

The *superior boundary* of the inguinal canal is formed by the lower margins of the internal oblique and transversalis muscles; these muscles are here intimately connected, particularly as we approach the pubes—towards the ilium they are partially separated. Beneath the former, the spermatic cord may be seen to emerge, as it descends; a few muscular fibres pass off along with it, which form the cremaster muscle. In some subjects a different arrangement exists; which is, that the spermatic cord passes out between the fibres of the internal oblique, some of which thus lying beneath the cord, this structure then escapes through a perfectly muscular opening. In this stage of the dissection, the extent of origin of the internal oblique muscle from Poupart's ligament may be observed; it in general arises from the external two-thirds and upper surface of the ligament, but is liable to much variety in this respect; the fibres descend inwards to terminate in the conjoined tendons.

The *cremaster muscle* arises not only from the lower margin of the internal oblique, but takes a few fibres from the transversalis muscle, from the neighbouring surface of Poupart's ligament, and from the pubes a little external to its tubercle; from these different origins, its fibres descend around the spermatic cord, but chiefly on its outer and anterior surface, forming a series of loops, the concavities of which are directed upwards, and are finally implanted into the outer surface of the tunica vaginalis, and into the scrotum, to the lower part of which they descend. Between the fibres of the muscle, the spermaticus supeficialis nerve, a branch from the ilio-scrotal descends, to be lost in the coverings of the testis.

Poupart's ligament bounds the inguinal canal *inferiorly*. Near the internal ring, the cord is about three-quarters of an inch above Poupart's ligament, but, as it descends, it gradually approaches the ligament, until at length it lies imbedded in its upper surface which is grooved for its protection.

The *anterior* boundaries of the inguinal canal have been already sufficiently described. Its *posterior* walls are the fascia transversalis and the conjoined tendons—this latter structure forms but a small portion of the posterior boundary of the canal—it lies behind the external abdominal ring. The *transversalis fascia* will be better seen immediately.

The *internal abdominal ring* is formed by the passage of the spermatic cord through the transversalis fascia; as the cord passes through the fascia it draws with it, as it were, a cellular prolongation from the fascia; this descends, and is gradually lost upon the cord; it is named the *fascia propria*, or *internal spout-like fascia*. It forms the fifth, or the immediate covering of the hernial sac, hence its name of *fascia propria*. It is well exposed by gently drawing downwards the spermatic cord.

The *internal abdominal ring* is a little nearer the ilium than the symphysis pubis, its inner margin being precisely midway between these two points. It lies nearly opposite the external iliac artery, and about one-half to three-quarters of an inch above Poupart's ligament. Directly *above* it the lower fibres of the transversalis muscle cross from their origin, from the external third or half of Poupart's ligament, almost transversely inwards, to become identified with the internal oblique muscle, and implanted with it into the conjoined tendons. Occasionally, as Mr. Guthrie has described, a few fibres of the transversalis muscle pass behind the cord. Along the *internal* and *inferior part* of the internal ring, and contained in the transversalis fascia, ascends the *epigastric artery*.

This artery arises from the external iliac, a little above Poupart's ligament; it at first descends, then curves upwards, winding around the cul de sac of the peritoneum, and then ascends inwards, enters the sheath of the rectus muscle, and terminates by anastomosing with the internal mammary artery. It lies about a *quarter of an inch* distant from the cord.

The epigastric artery is accompanied by one or two veins; if by one, the vein lies to its inner or pubal side; if by two, the artery lies between the veins.

The following measurements of the parts connected with hernia are given by Sir A. Cooper.

From symphysis pubis to the	M. ins.	F. ins.
anterior superior spinous process of the ilium	$5\frac{3}{4}$	6
tuberosity of the pubes	$1\frac{1}{8}$	$1\frac{3}{8}$
inner margin of the lower opening of the abdominal canal	$0\frac{7}{8}$	1
inner edge of the internal abdominal ring	3	$3\frac{1}{4}$
to the middle of the iliac artery	$3\frac{1}{4}$	$3\frac{3}{8}$
iliac vein	$2\frac{5}{8}$	$2\frac{3}{4}$
origin of the epigastric artery	3	$3\frac{1}{4}$
course of the epigastric artery on the inner side of the internal abdominal ring	$2\frac{3}{4}$	$2\frac{7}{8}$
middle of lunated edge of the fascia lata	$3\frac{3}{4}$	$2\frac{3}{4}$

	M.	F.
From the anterior edge of the crural arch to the saphena vein . . .	ins. 1	ins. 1¼
————— symphysis pubis to the middle of the crural ring . . .	2¼	2⅜

The *transversalis fascia* is a layer of condensed cellular tissue or fascia, situated between the transversalis muscle and the peritoneum, and particularly well marked, *inferiorly*, where the internal oblique and transversalis muscles are deficient.

Superiorly, the transversalis fascia is gradually lost in the loose cellular tissue which connects the peritoneum on its anterior and posterior surfaces to the abdominal parietes, but *inferiorly* becomes strong and well marked. Here, on the *outer* side, it is attached to the inner edge of the crest of the ilium, more *internally* to Poupart's ligament, behind which it unites with the fascia iliac to prevent the protrusion of hernia beneath this ligament; still more internally, or opposite the fermoral vessels, it descends into the thigh, forming the anterior layer of the canal in which these vessels are lodged; still nearer the pubis, it is attached to the conjoined tendons, and the outer edge of the rectus muscle, a layer of it passing also behind this muscle to be continuous with that of the opposite side.

At the point where the spermatic cord perforates the transversalis fascia, namely, the internal abdominal ring, the fascia frequently presents a well-defined semilunar border, which bounds the ring to its inner side; in other cases it terminates gradually in this direction; in all instances a prolongation is derived from it which descends on the spermatic cord, and forms the fascia propria, or fascia spermatica, of Sir A. Cooper. Some anatomists describe the transversalis fascia as being composed of two layers, between which the epigastric artery ascends, one of these being attached to the edge of the rectus muscle, the other passing behind the muscle to the opposite side. The transversalis fascia is of use in strengthening the abdominal parietes inferiorly, where a deficiency exists in the internal oblique and transversalis muscles. From its connexion with the transversalis and rectus muscles, it is made tense by the action of these, and is thus rendered still more capable of opposing the protrusion of a hernia.

As the hernia is first protruded against the internal abdominal ring, it pushes before it the peritoneum, which forms the hernial sac, emerging through the ring it descends in front of the spermatic cord, and here receives a covering from the fascia propria; descending still further, it escapes beneath the lower margin, or between the fibres of the internal oblique muscle, here it enters the sheath of the cremaster muscle, and continues its course to the external abdominal ring; it now, turning forwards, passes through this opening, where it becomes invested by the external spout-like fascia; it now descends into the scrotum, covered still further by Scarpa's fascia, the superficial fascia, and the integuments. It thus is covered by the *integument*, the *superficial fascia*, *Scarpa's fascia*, the *external spout-like fascia*, the *sheath of the cremaster muscle*, the *fascia propria*, and the *peritoneum*, forming the hernial sac.

As a hernia descends into the scrotum, beneath the coverings of the spermatic cord, it finally arrives immediately above the testis, in which position it is retained by the attachment of the spermatic coverings to the upper and back part of this organ, where its vessels and nerves enter.

Now the principal points to be attended to, in the anatomy of oblique inguinal hernia, are—1st, its *coverings;* 2d, the part of the canal where stricture most frequently occurs; 3d, its relation to the epigastric artery, to the spermatic cord, and the testis.

Oblique inguinal hernia in the female occurs much less frequently than in the male, owing to the smaller size of the inguinal canal. It descends into the labium and is in general small. It is almost always reducible. Should it become strangulated, the operation will not differ essentially from that recommended for the male. The lower part of the peritoneal bag, according to Sir A. Cooper, contains only water. The incisions should not be prolonged much into the labium.

Direct inguinal hernia.—From the deficiency in the external oblique tendon, described in oblique inguinal hernia under the name of the external abdominal ring, it follows that a hernia may be protruded directly forwards from the abdomen through this aperture, without transversing the inguinal canal; this will therefore form the direct inguinal hernia.

Fortunately nature has so protected this opening posteriorly, that this species is comparatively rare. The parts that oppose its formation lie behind the ring; they are, the triangular ligament, or Colles' fascia, the conjoined tendons, the edge of the rectus muscle, and the transversalis fascia, attached to the outer edge of this muscle and to the conjoined tendons.

These barriers are not, however, in all cases sufficient. As a direct inguinal hernia makes its way forwards through the external ring into the scrotum, it pushes before it the peritoneum which forms the hernial sac; it next either bursts through, or carries with it, the transversalis fascia, and then passing to the outer edge of the rectus muscle and the conjoined tendons, it escapes through the external ring, and is here covered by the external spout-like fascia, the superficial fascia, and the integument. It is evident that the hernia can receive no covering (or at least a very partial one) from the cremaster muscle, nor from the fascia propria of oblique inguinal hernia, as it does not descend through the inguinal canal.* The

---

*Sir A. Cooper states that it receives a covering from the transversalis tendon and fascia. By the former we consider him to mean an expansion from the transversalis fascia, where it is attached to the conjoined tendons.

epigastric artery lies to its outer side. As the hernia emerges from the external ring, it lies *internal*, and a little *posterior*, to the spermatic cord.

Direct thus differs essentially from *oblique* inguinal hernia in its *coverings*, and in its *relations* to the *epigastric artery* and *spermatic* cord.

If we examine the lower part of the parietes of the abdomen on their inner surface, by raising the intestines from the cavity of the pelvis, leaving the peritoneum in situ, we shall there see two depressions, or fossæ, on each side, which seems to invite, as it were, to the protrusion of a hernial tumour. Between these three projecting cords may be noticed; one of these is situated on the mesial line, and is formed by the urachus, which ascends from the superior fundus of the bladder to the umbilicus; the remaining two are situated one on each side. They are formed by the degenerated umbilical arteries, which ascend from the lateral surfaces of the bladder, converging to the umbilicus.

These depressions are named the *internal* and *external inguinal fossæ* or *pouches*. The *internal inguinal fossa* lies between the urachus and the degenerated umbilical artery, the first of which separates it from its fellow of the opposite side. It is triangular in shape, the base inferiorly being formed by the crest of the pubes. It lies opposite to the *external ring*, and thus leads to the formation of *direct inguinal hernia*.

The *external inguinal fossa* lies to the outer side of the umbilical artery; it lies superior and external to the other, and is somewhat larger; it leads to the *internal ring*, and consequently to the formation of an *oblique inguinal hernia*.

The projecting cords, formed by the urachus and umbilical arteries, contribute more to the formation of hernia than the inguinal fossæ; as they, when the intestines are propelled against them by the action of the abdominal muscles, give to the propelling force, and consequently to the viscera, a direction forwards towards the fossæ.

Femoral hernia.—This form of hernia is more frequent in the female than in the male, in consequence of the greater breadth of the pelvis in the former. It is sometimes named crural hernia, being called from its making its appearance, when first noticed, in the thigh; in the upper, inner, and anterior part of which region it descends beneath Poupart's ligament, to the inner side of the femoral vessels.

Beneath the skin is the *superficial fascia*. This is the common subcutaneous cellular tissue met with in every part of the body beneath the integuments; it is here well marked, and is frequently loaded with fat. On all sides it is continuous with the surrounding cellular tissue. Externally a few transverse fibres may be occasionally met with in it. No anatomist, so far as we are aware of, has as yet exercised his ingenuity in the vain-glorious task of dividing it into any definite number of layers. It forms the most superficial covering of the femoral hernia.

In the superficial fascia we notice several arteries, veins, nerves, and lymphatic glands. The *arteries* are branches of the external circumflexa ilii, external pudic, and superficial epigastric arteries; they arise from the femoral artery a little below Poupart's ligament, perforate the cribriform fascia, and are lost in the superficial fascia and integuments. The external circumflexa ilii branches pass outwards behind the ilium; the external pudic to the organs of generation; whilst the superficial epigastric ascends inwards over Poupart's ligament, to be lost in the parietes of the abdomen. The external pudic and superficial epigastric are frequently wounded in the operation for strangulated femoral hernia, and may require the ligature.

The *veins* are branches from the surrounding integuments, and the *great* or *internal saphena* vein. This large vein commences at the inner side of the foot and leg, ascends along the internal surface of the thigh, curves outwards at its upper part, and at about an inch and a half distance from Poupart's ligament, perforates the cribriform fascia, through the saphenic opening, to join the femoral vein.

In some cases a second large vein ascends along the anterior surface of the thigh, which either terminates in the saphena or in the femoral vein. In operating for strangulated femoral hernia, as well as in other operations in this region, the saphena vein is much exposed. It may be avoided by not prolonging our incisions too much inwards. Not unfrequently it becomes varicose, and may be mistaken for a femoral hernia.

The *nerves* met with in the superficial fascia are small branches derived principally from the anterior crural.

The *lymphatic glands* compose the interior set of inguinal glands. They are arranged in the vertical direction, and consist of two sets, the *superficial* and *deep*. The *superficial* are three or four in number, and are enclosed in a capsule of the superficial fascia. The *deep* are two or three in number, they accompany the femoral vessels; one is almost constantly lodged in the femoral ring.

On raising the superficial fascia, the *fascia lata* is brought into view. This is a dense layer of fibrous structure which invests the muscles of the thigh. *Inferiorly* it is attached to the tendons and ligaments about the knee-joint; from this it ascends, fórming sheaths for the different muscles, and sending in processes between them. Having arrived at the upper and anterior part of the thigh (where it is connected with the anatomy of femoral hernia,) it divides into three portions, the *iliac, pectineal, and cribriform fasciæ*.

The *iliac* (so called from its connexion with the os ilii,) or *external* portion of the fascia lata, is the strongest; posteriorly it is attached to the crest of the ilium; in front we find it closely attached to the lower margin of Poupart's ligament. As we trace it inwards towards the pubes, it terminates in an elongated process, which passes in front of the femoral vessels, gets to their inner side, and is implanted into the front of the pectineal portion of the fascia lata. Some of its fibres are here reflected behind Poupart's ligament, they become continuous with the base of Gimbernaut's ligament, and are inserted with it into the ilio-pectineal line. The *superior* edge of this prolonged portion of the fascia lata is attached to Poupart's ligament—its *inferior* margin is free and concave, the concavity looking downwards and inwards; this portion of it is named Hey's ligament, or the *falciform process* of the fascia lata, whilst its point of attachment to the pubic portion of the fascia lata is called Colles' ligament.

A short distance below Hey's ligament the iliac portion of the fascia lata is gradually prolonged into the cribriform fascia, but not unfrequently presents, in the dissected state, a well-defined edge at the outer side of the femoral artery. Still more *inferiorly*, the iliac portion of the fascia lata becomes continuous with the pectineal portion in front of the femoral vessels; they here conjoined form a well-defined semilunar margin, the concavity of which is turned upwards, and forms *Burns' ligament*. This margin of the fascia lata, on a careful examination, will be found to be reflected backwards on the anterior surface of the femoral vessels, and so intimately united with the sheath, as to prevent altogether the descent of a hernia beneath it.

The *pectineal* or *pubic* or *internal* portion of the fascia lata lies in front of the pectineus muscle—*internally* it is attached to the symphysis pubis, where it lines the gracilis and adductor muscles; as we trace it outwards it passes in front of the pectineus muscle, gets behind the femoral vessels, at the outer edge of which it meets with the tendons of the psoas magnus and iliacus internus muscles: it here divides into two laminæ, one of which passes forwards, to be attached to the posterior surface of the external or iliac portion of the fascia lata, whilst the other passes backwards to be attached to the capsular ligament of the hip-joint.

Superiorly the pectineal fascia ascends in front of the pectineus muscle, to be inserted in the ilio-pectineal line, where it becomes continuous with the fascia iliaca, and with Gimbernaut's ligament. *Inferiorly* it becomes continuous with the iliac portion of the fascia lata, to form Burn's ligament.

Femoral hernia, as it descends, rests on the pectineal fascia. Between these two portions of the fascia lata, and immediately in front of the femoral vessels, an oval-shaped space exists, which is covered over by a thin layer of cellular substance, named the *cribriform fascia*, or *middle* portion of the fascia lata. This is in general described as a process of the fascia lata, with which indeed it is perfectly continuous at the circumference of the opening; at the same time, it is so intimately connected to the superficial fascia, being in fact identified with it, that it may be, with equal propriety, described as being derived from this fascia.

The cribriform fascia is perforated, as its name implies, by a number of foramina, which transmit numerous veins and nerves, passing between the superficial and deeper parts. One larger than the rest is formed by the saphena vein dipping in to join the femoral; this is the saphenic opening; through it a femoral hernia most frequently passes forward beneath the superficial fascia; it is bounded inferiorly by Burn's ligament. From the little resistance the cribriform fascia affords, a femoral hernia soon makes its way through it; this it effects either by bursting through the fascia, or dilating one of the openings in it, most frequently the saphenic opening.

Having examined the parts on the anterior surface of the thigh, we proceed to ascertain by what means a hernia is enabled to escape from the cavity of the abdomen, so as to descend behind the fascia lata. For this purpose draw the peritoneum and viscera out of the iliac fossa, and examine the parts which descend beneath Poupart's ligament. As this ligament stretches across the brim of the pelvis, (between the anterior superior spinous process of the ilium and the tubercle of the pubes) it leaves a large space (the crural arch) between it and the bone, by means of which several structures are transmitted from the pelvis to the thigh, or in the opposite direction. Thus between the spine of the ilium, the inguino-cutaneous nerve descends, internal to this the psoas magnus and iliacus internus muscles, and between them the anterior crural nerve, descend in a deep fossa; still more internally the femoral vessels pass. On the inner side of these the absorbent vessels from the lower extremity ascend through a well-marked opening (the femoral ring.)

So many parts thus passing beneath Poupart's ligament, nearly obliterate the space between it and the bone, so that the descent of a femoral hernia is, by these means, at least partially prevented. As, however, many interstices must exist between these parts, it becomes necessary that some additional structure should exist, in order to preserve the integrity of the abdominal cavity. This structure does exist, and is formed by the junction of the transversalis fascia and the fascia iliaca behind Poupart's ligament.

The *fascia transversalis* descends from the inguinal canal, passes backwards, and behind Poupart's ligament meets with the *fascia iliaca*.

This is a dense layer of fascia, which invest the anterior surface of the iliacus internus muscle; *externally* and *superiorly* it is firmly attached to the inner lip of the crest of the ilium. Tracing it *inwards*, it lines the muscle, passes behind the external iliac vessels, sending off at the same time an expansion, which passes in front of these vessels, binds them down, and forms the fascia propria of the external iliac artery, first described by Mr. Abernethy. At the inner side of the iliac vessels, the fascia iliaca becomes attached to the ilio-pectineal line, whence it descends into the pelvis, under the name of the pelvic fascia. Near the pubes the fascia iliaca becomes continuous with the pectineal portion of the fascia lata, and Gimbernaut's ligament. *Inferiorly*, or towards the femoral region, the fascia iliaca curves forwards, and meets the transversalis fascia immediately behind Poupart's ligament, where the two fasciæ, united, form a dense whitish tendinous line, which indicating the course of the internal circumflexa ilii vessels, extends from the spine of the ilium to the outer edge of the femoral artery, and thus precludes the possibility of a hernial tumour descending between these points.

At the *outer* side of the femoral artery the fascia transversalis and the fascia iliaca separate; the fascia transversalis descends into the thigh in front of the femoral vessels, whilst the fascia iliaca descends behind them, thus enclosing these vessels in a distinct well-marked sheath. This, the sheath of the femoral vessels, is of a funnel shape; the base superiorly, the apex inferiorly, is gradually lost in the cellular coverings of the femoral vessels in the thigh.

The femoral artery and vein are not in close contact whilst contained in the sheath, but are separated by a septum or partition, which passes from its anterior to its posterior wall; and which, thus opposing the separation of these walls, prevents the descent of a hernia between the femoral vessels and the fascia transversalis, or anterior wall of the femoral sheath. In addition to this, the fascia transversalis is connected by cellular tissue to the cellular covering of the vessels, so as to assist materially in preventing a descent of a hernia in this direction.

On the *inner* side of the femoral vessels, we have just stated that a large aperture exists for the transmission of the absorbent vessels from the lower extremity. Here the provisions against the descent of a hernial tumour are but trifling, and here therefore it is that a femoral hernia first makes its way from the cavity of the abdomen. This opening is named the *femoral ring*, and should be attentively studied. It is as yet concealed from view by a delicate layer of cellular tissue, which may be described as the termination of the fascia transversalis, on the inner side of the femoral vessels. This forms the *fascia propria* or femoral hernia, inasmuch as a hernial sac, protruding through the femoral ring, carries the fascia before it, and thus receives from it an immediate investment in a fascia propria. Although in the natural state the fascia propria is thin and delicate, yet in old cases of hernia it becomes thickened and condensed, so as to present a membranous appearance. In some cases it may be burst through, so that no fascia propria will then exist. Mr. Guthrie has recorded an instance of this.

On removing the fascia propria, the *femoral ring* will be at once exposed. It is triangular in shape, the *base* externally at the femoral vein, the *apex* internally at Gimbernaut's ligament; it is bounded in *front* by Poupart's ligament, and the reflected portion of the falciform process of the fascia lata, *behind* by the horizontal ramus of the pubes, covered by the pectineus muscle and the pectineal fascia. The spermatic cord, or ligamentum teres, lies a little above and in front of it, and the epigastric artery curves to its outer side. In it we often meet with a lymphatic gland.

Here, then, is the unprotected part of the abdomen, through which a femoral hernia descends into the thigh. As the hernia escapes through the ring, it pushes before it the peritoneum, which forms the hernial sac, as also the fascia propria, which, together with the lymphatic gland in the ring, affords but a feeble resistance; the latter is soon pushed aside, the former becomes one of the coverings of the hernia, which, descending into the thigh beneath Poupart's ligament, rests on the pectineal fascia, behind the falciform process of the fascia lata. It soon arrives opposite the cribriform fascia. Its further descent is prevented by the convexity of the saphena vein, by the attachment of Burn's ligament to the front of the femoral vessels, and by the motions of the thigh on the pelvis. Meeting with little resistance from the cribriform fascia, it changes its course, turns *forwards*, bursting through the fascia, or dilating one of the apertures in it, most frequently the saphenic opening, and lies beneath the superficial fascia. It soon changes its course a second time; it now turns *upwards*, ascends over Poupart's ligament, and finally rests above this ligament, on Scarpia's fascia of inguinal hernia, which alone separates it from the tendon of the external oblique muscle. It is here then covered by the *integument, superficial fascia, fascia propria*, and *hernial sac*.

We have stated that the femoral ring is bounded internally by *Gimbernaut's ligament*. This is the third insertion of Poupart's ligament, from the pubic extremity of which it extends obliquely inwards and backwards, to be inserted in the ilio-pectineal line; it is triangular in shape, the apex turned inwards towards the pubes; it base is directed outwards, is semilunar, and forms the inner boundary of the femoral ring; one edge is turned forwards, and is attached to Poupart's ligament; the other is directed backwards and inwards, and is implanted into the iliopectineal line. Gimbernaut's ligament is of use in preventing the descent of a femoral hernia on the inner side of the femoral ring.

The femoral ring possesses in many subjects a relation which should not be overlooked; it is the *obturator artery*. This vessel in general arises from the internal iliac artery. Where this is the case, it can have no relation to a femoral hernia; but in numerous instances it will be found to arise from the epigastric, close to the origin of this vessel from the external iliac artery.

Now as its ultimate distribution is to the parts in

the neighborhood of the obturator foramen; it necessarily follows, when such is its origin, that in order to arrive at this foramen, it must cross the femoral ring, and consequently a femoral hernia, if such should exist. As it proceeds to the foramen, it may either pass directly to it along the outer side of the neck of the hernia; or it may first pass along its posterior surface, and then enter the foramen; or it may wind along the anterior surface of the neck of the sac, and then descend to the inner side.

Section V.

The Urinary Organs.

These organs consist of the kidneys, ureters, bladder, and urethra.

The KIDNEYS are situated one on each side, in the right and left lumbar regions, behind the colon, in front of the quadratus lumborum musc., and correspond to the last two dorsal and first two lumbar vertebræ, reaching from the last rib to the crest of the ilium; the *anterior surface* is convex, the *posterior* flattened; their extremities are convex, the upper being a little the larger, the *ext.* margin is convex, the *int.* concave and notched for the passage of the vessels and ureters; the right touches the liver superiorly, the cæcum inferiorly; the left the spleen above, the sigmoid flexure of the colon below; each kidney is surmounted by the suprarenal capsule. The kidney has but one coat, the fibrous; it is strong and well marked.

The *internal structure* of the kidney is composed of the *secretory* or *cortical* substance, and the *excretory* or *tubular* portion. The *secretory* or *cortical* substance forms its greater portion and occupies its outer part; it is arranged in conical masses, *the cones*, about fifteen in number, which terminate in pointed extremities, the *mammillary processes;* these are about twelve or thirteen in number, and are surrounded by the *calyces.*

The *tubuli uriniferi* proceed from the cortical substance, to which they give a striated appearance; they open on the mammillary processes.

The calyces, six or eight in number, form conical cups, into which the urine is dropped from the mammillary processes; they unite into three tubes, the *infundibula,* which uniting form a funnel-shaped tube, the *pelvis* of the kidney: this terminates in the ureter.

Each kidney is supplied with blood from the renal art.; the right is larger than the left, its blood is returned by the renal vein, which opens on each side into the inf. vena cava. Its nerves are derived from the renal plexus formed from the lesser splanchnic, sympathetic, and solar plexus.

The *ureter,* the excretory duct of the kidney, proceeds from the pelvis of the kidney downwards and inwards, across the psoas musc., and iliac art. and behind the peritoneum; it then sinks backwards into the pelvis, turns forwards, and comes in contact with the inferior surface of the bladder; it runs forwards and inwards between the muscular and mucous coats of this viscus for a short distance, and finally opens into it at the posterior angle of the trigone vesicale. The ureter is crossed superiorly by the spermatic vessels, near its termination by the vas deferens in the male, and by the Fallopian tube and broad ligament in the female. The ureter is composed of fibro-mucous membrane; the fibrous coat is well marked, its cavity is smallest at its termination, largest at its commencement.

The *suprarenal capsules* are two yellowish bodies placed on the upper ext. of each kidney; they are largest in the fœtus, at which period they exceed the kidneys in size; they degenerate into cellular tissue in the adult.

9*

The *bladder* is lodged in the cavity of the pelvis, behind the pubes, in front of and above the rectum, and between the levatores ani muscles; pyramidal in shape, its apex rounded, is directed upwards and forwards, its base, directed downwards and backwards, rests on the lower extremity of the rectum, from which it is separated by the prostate gland, vesiculæ seminales, and vasa deferentia: its *anterior* surface, somewhat flattened, is in contact with the pubes, its *posterior* with the rectum and small intestines. In the female the uterus intervenes between the bladder and rectum.

The bladder is composed of four coats, *serous, muscular, fibrous,* and *mucous;* the *serous* coat is derived from the peritoneum, it is only a partial covering; the parts covered are the posterior half of superior fundus or apex, the posterior surface and the anterior portion of the lateral surfaces, the rest is uncovered by peritoneum; the *muscular* coat consists of three sets of fibres; the *longitudinal,* best marked, are situated on the anterior and posterior surfaces, they unite at the superior fundus at the attachment of the urachus, and inferiorly are lost in the neck of the bladder; the *oblique* fibres are best marked on the lateral surfaces, the *circular* surround the neck of the bladder.

The *fibrous* coat of the bladder is well marked; the *mucous* coat lines the interior, it is of an irregular pinkish colour, and is generally thrown into folds in the collapsed state; on it may be observed the orifices of the ureters; immediately behind the neck of the bladder is a triangular space, pale and smooth, the *trigone vesicale;* its apex is at the orifice of the uretha, its base is formed by a line drawn from the orifice of one ureter to that of the other, its sides formed by lines drawn from them forwards to the apex, beneath them some muscular fibres are described. The ligaments of the bladder are the false and true; the false are folds of the peritoneum, and are five in number,

THE URETHRA.

viz., two posterior, two lateral, and one anterior; the true are folds of the pelvic fascia, and are four in number, viz., two *lateral*, which pass off from the inner surface of the levator ani on the lateral surface of the bladder, and two *anterior*, which run from the neck to the back of the os pubis.

The bladder is supplied with blood chiefly by the vesical branches of the int. iliac, its nerves are derived from the hypogastric plexus, and consists of spinal and sympathetic filaments.

The *urethra* extends from the neck of the bladder to the orifice on the glans penis, about nine inches in length, it is divided into three portions, the prostate, membranous, and spongy. The *prostatic portion* is surrounded by the prostate gland; about an inch and a half in length, it runs downwards and forwards, its cavity is wider in the centre than at its extremities; on its lower surface is a projection of the mucous membrane, the *caput gallinaginis* or *verumontanum*, on the anterior margin of this is a depression, the *sinus pocularis*, and on each side of this a small orifice, the opening of the common ejaculatory duct, at the side of the verumontanum is a fossa, the *prostatic sinus*, on which a number of the prostatic ducts open.

The *membranous portion* of the urethra lies immediately beneath the sub-pubic arch, from which it is separated by the sub-pubic ligament, the dorsal veins of the penis, and the apex of the triangular ligament; about three-quarters of an inch in length, it is slightly curved; its cavity is smooth and small in size, its external covering is a dense fibrous membrane. The *spongy portion* of the urethra is that surrounded by the corpus spongiosum urethræ, it forms the remainder of the canal; it commences posteriorly by a small dilatation, the sinus of the bulb, and terminates at the ext. orifice, the spongy portion dilates a little again behind the orifice forming the *fossa navicularis*, and then contracts to form the orifice—a slit-like aperture on the glans penis. In the interior of this portion of the canal are numerous lacunæ, one larger than the rest, *locuna magna*, is situated near the fossa navicularis. The urethra in the female is short, about one inch and a-half in length, and opens externally beneath the clitoris between the nymphæ minores, and immediately above the orifice of the vagina.

Section VI.

Male Organs of Generation.

These consist of the testes, vasa deferentia, vesiculæ seminales, and penis.

The *testes* are enclosed after birth in the scrotum, before this period they are placed a little beneath the kidneys, whence they descend into the scrotum.

The *scrotum* is formed by a continuation of the common skin, the dartos, superficial fascia, tunica communis and tunica vaginalis. The skin of the scrotum is of a darkish brown and corrugated; it presents on the mesial line the raphè; the *dartos* is a thin layer of muscular fibres seldom well marked; the superficial fascia is continuous with that of the surrounding parts, it contains no adipose substance; the *tunica communis* is a layer of fibrous membrane derived superiorly from the cremaster muscle, and continuous with the fascia investing the spermatic cord above; the *tunica vaginalis* is that layer of this membrane which passes from off the testis to the inner surface of the scrotum; it thus forms a smooth cavity in which the testis is lodged. The scrotal or testal cavity on each side is distinct from the opposite one, being separated by the *septum scroti*, which is formed by a reflection inwards of the dartos, and the superficial fascia.

The *testes* are contained in the scrotal cavities. Each testis is of an oval shape, the long axis being directed downward, backwards, and a little inwards, the outer surface is convex, the inner somewhat flattened, the lower extremity is the smaller, the upper is capped by the head of the epididymis. It is composed of three coats, a serous, fibrous, and vascular. The *serous* coat is a portion of the tunica vaginalis; this, originally a reflexion of the peritoneum, forms in the adult a distinct shut sac, lining the inner surface of the scrotum (tunica vag. scroti) and the outer surface of the testis (tunica vag. testis;) it is partial, and does not cover the testis on its back part, where the epididymis is attached and the vessels enter.

The *fibrous* coat is the *tunica albuginea*. It is very dense, and of a whitish colour, the fibres interlace with each other, and send processes through the substance of the testis, which, as they pass backwards, form an imperfect septum, mediastinum testis, and terminate in and form the *corpus Highmorianum*.

Tunica vasculosa lines the internal surface of the preceding; it is formed by a delicate network of vessels.

The *proper substance* of the testis consists of an infinite number of small tubes, *tubuli seminiferi*, which form a grayish pulpy mass; they terminate in about twenty larger vessels which take a straight course, hence called *tubuli recti*, towards the back part of the testis; they here form a network with the vessels and nerves of the testis forming the *rete testis*, lying between the layers of the mediastinum testis; from the upper and back part of this eight or ten vessels proceed, called *vasa efferentia*, or *coni vasculosi*, and terminate in the head of the epididymis, forming one tube, the *vas deferens;* this is convoluted in a remarkable manner, and forms the *epididymus ;* this emerges from the globus minor, ascends along the inner edge of the epididymis, and forms part of the spermatic cords. The *vas deferens* ascends with this, passes along the inguinal canal, winds inwards at the internal abd. ring, and then descends into the pelvis to the inner side of the hpyogastric art. and ureter, and reaches the inferior fundus of the bladder; it here runs forwards and inwards along the inner side of the vesicula seminalis, perforates the prostate gland and unites with the duct of the vesicula seminalis to form the *common ejaculatory duct,* and terminates on the verumontanum; the vas deferens lies on the posterior part of the cord, where it may be distinguished by its whipcord-like feel. Its structure is fibromucous; the fibrous coat is very dense; the canal is extremely small, being not larger than a bristle.

The *spermatic cord* is formed of the vas deferens, the spermatic art., veins, and nerves; it extends from the epididymis to the int. ring, where its component parts separate. The vas deferens proceeds as described. The spermatic art. ascends to its origin from the renal, the veins to the renal veins, and the nerves to the renal plexus and sympathetic nerve. The coverings of the cord have been described with hernia, the remains of the peritoneum covering the cord are sometimes described as the tunica vaginalis of the cord.

The *vesiculæ seminales* are situated one on each side of the inferior fundus of the bladder to the outer side of the vas deferens; pyriform in shape, the larger extremity is directed backwards and outwards, the smaller forwards and inwards, terminates in a small duct which unites with the vas deferens, and terminates with it as just described. The structure of the vesiculæ seminales is fibro-mucous, its interior presenting a number of cells, with a central longitudinal canal.

The *prostate gland* surrounds the neck of the bladder and first portion of the urethra, which passes near to its upper surface, one-third of the gland being above, and two-thirds beneath this canal. The prostate gland is covered by the ascending layer of the triangular ligament of the urethra, and has attached to it anteriorly the anterior true ligaments of the bladder which connect it to the pubic arch.

The *prostate gland* is chestnut-shaped, the base posteriorly, the apex anteriorly; it consists of three lobes, two lateral and a middle lobe; the lateral are separated by a slight depression on its under surface, the middle lobe lies in and a little above the centre of the sulcus at its posterior extremity. This gland is composed of a number of mucous follicles, its ducts open into the prostatic portion of the urethra, the greater number on the sinus poculosus.

The *penis* arises by two roots, the crura penis, one on each side, from the rami of the ischium and pubes; these are conical in shape, pass forwards, converging, and unite to form the *corpus cavernosum penis*, which forms the body of the penis as far forwards as the glans: as the corpus cavernosum is partially divided by an imperfect *septum*, it may be considered as consisting of two lateral portions. This body is covered exteriorly by a dense fibrous structure, which forms the septum, within which is a peculiar erectile tissue composed chiefly of arteries. The penis presents on its *upper* surface a slight depression, in which are lodged the dorsal vessels and nerves; on its *under* surface is a larger depression, giving lodgement to the corpus spongiosum urethræ; it is covered by a delicate thin skin, which anteriorly forms the prepuce, connected to the under surface of the glans by the *frænum preputii;* where this is reflected on the corona glandis, a number of sebaceous follicles, *glandulæ Tysoni*, are found.

The *glans penis* is formed altogether by the corpus spongiosum urethræ. The penis is supplied with blood by its dorsal and cavernous vessels; branches of the int. pubic; its veins pass backward beneath the sub-pubic arch, and form a network around the prostate gland, and terminate in the int. iliac; its nerves are derived from the int. pubic. The corpus spongiosum is supplied by the the art. of the bulb, a branch also of the int. pubic; its structure is also erectile, and seems to be chiefly composed of veins.

Section VII.

The Perinæum

is a diamond-shaped space situated between the thighs on each side, the scrotum and sub-pubic angle in front, and the os coccygis behind. In this is included the anal region, properly so called, which occupies the posterior triangle, whilst the proper periuæum occupies the anterior. The lateral boundaries of the perinæum are on each side the rami of the ischium and pubes, the tuber ischii, and the edge of the glutæus maximus muscle, and, more deeply, the great sacrosciatic ligament.

The integuments of this region are thin, of a brownish color, and present on the mesial line a fold termed the *raphé*. This commences at the anus, and may be traced as far forwards along the mesial line of the scrotum as the root of the penis. Beneath the integuments is a quantity of loose cellular tissue, the superficial fascia of the perinæum, continuous on either side with the superficial fascia of the thighs, in front with that of the scrotum, and posteriorly with the adipose tissue filling up the large space at the side of the rectum.

Beneath the superficial fascia is the *perinæal fascia;* this is much more dense than the preceding, and is firmly attached on each side to the rami of the ischium and pubes; anteriorly it is lost on the scrotum, and posteriorly is continuous with the adipose tissue surrounding the rectum. Beneath this lie

The Muscles of the Perinæum.

Transversus Perinæi.—*Or.* from the inner surface of the tuber. ischii, above the erector penis m.

Ins. into the *central point* between the accelerator urinæ and sphincter ani m..·. these two muscles correspond to the transverse line just referred to..·. one or two more m. (T. P. Alteri) sometimes pass upwards and inwards to the accelerator urinæ m. *Use*, to fix the central point of the perinæum.

Accelerator urinæ.—*Or.* 1, the *central point* and two inches of the fibrous raphé running from it along the corpus spong.; 2, from the post. part of the triangular ligament. *Ins.* 1, the posterior fibres into the inner side of the crus penis; 2, the middle by distinct aponeurotic fibres into the groove between the two corp. cavernosa, commencing at the junction of the two crura; 3, the anterior pass upwards and forwards over the corp. cavern., and are *ins.* into the suspensor lig. of the penis.·. The ant. fibres compress the dorsal veins of the penis. (Houston.) *Use*, to compress the bulb and corpus spongiosum urethræ, and to expel their contents, whether urine or semen; hence called *ejaculatores seminis*.

Erector penis—*Or.* 1, fleshy and tendinous from the inner edge of the tuberosity of the ischium, and root of the crus penis. *Ins.* by a tendinous expansion into the fibrous membrane of the corpus cavernosum. *Use*, to erect the penis by forcing the blood forwards.

The whole of these muscles being removed, the crura penis detached, and the urethra cut through in front of the bulb, and reflected backwards, the *deep perinæal faschia* or *triangular ligament* of the urethra is exposed. This is a dense fibrous layer closing up the anterior triangle of the perinæum. On each side it is firmly attached to the rami of the ischium and pubes, and tuber ischii, and is here continuous with the outer layer of the pelvic fascia. Its apex, turned forwards, passes above the urethra, splits into two layers, enclosing the sub-pubic ligament, and lost on the surface of the pubes; its base, turned backwards, is crescentic to accommodate the concavity of the rectum, and is lost on the surfaces of this intestine.

The triangular lig. is perforated near the centre by the memb. portion of the urethra, which passes through it from above downwards and forwards. As the urethra passes through, the ligament sends off two layers, the *ascending* and *descending.* The *ascending* passes upwards and backwards, and is lost on the capsule of the prostate gland; the *descending* passes downwards and forwards, and is lost on the fibrous covering of the corpus spongiosum urethræ. In the angle between this and the bulb lie two small glands, *Cowper's glands*, which open by small ducts on the interior of the bulb.

The posterior triangle of the perinæum is occupied by the anus and lower ext. of the rectum. Around the anus is the

Sphincter ani muscle.—*Or.* from the extremity of the coccyx; its fibres surround the anus, and are *Ins.* into the central point of the perinæum. *Use*, to close the anus.

At each side of the rectum is a large space occupied by a quantity of adipose tissue; on removing this, the following muscle is exposed—

Levator ani.—*Or.* 1, from the posterior part of the symphysis pubis, below the true ligaments of the bladder; 2, from the obturator fascia, and from the ilium above the thyroid foramen, by means of the pelvic fascia; 3, from the inner surface and spine of the ischium, the fires converge, and are *Ins.*, the anterior fibres into the central point of the perinæum and fore part of the rectum, the middle into the side of the rectum, the posterior into the back part of the rectum and the sides of the os coccygis. *Use*, to raise and draw forwards the lower ext. of the rectum; also to assist the expulsion of the urine and semen.

The anterior fibres of this muscle are sometimes

described as *Wilson's muscles;* a few of its fibres are also described as *Guthrie's muscles.*

Behind the levator ani is the

Coccygeus.—*Or.* narrow from the inner surface of the spine of the ischium. *Ins.* into the side of the coccyx.

The *arteries* met with in the perinæum are the transversus perinæi, perinæal, and the art. of the bulb, branches of the int. pudic. Around the anus are the inferior hemorrhoidal arteries. The *nerves* proceed from the int. pubic and sciatic.

Section VIII.

The Female Organs of Generation

are divided into the external and internal. The *external* are the *mons veneris, vulva, labia, clitoris, nymphæ,* and *vagina.*

The *mons veneris* is the eminence on the anterior surface of the pubes. The *vulva* is the slit-like aperture between the labia. The labia externa, or majora, are the folds of integument on each side of the vulva; they commence at the mons veneris, and terminate posteriorly in the commissure or fourchette, behind which is the fossa navicularis; their structure is composed of skin externally, mucous membrane internally, enclosing some cellular vascular tissue. The *nymphæ*, or *labia minora* descend outwards from the prepuce of the clitoris, and are lost about the centre of the vulva on the labia majora; the orifice of the urethra lies between them. The clitoris is analogous to the male penis; it is covered anteriorly by an irregular prepuce, continuous inferiorly with the nymphæ. It contains no canal. M. Huguier has recently described two glands, situated one on each side and within the vagina, which opens by ducts near the margin of the hymen.

The *vagina* is the canal leading from the vulva upwards and backwards to the uterus; it is slightly curved, the concavity towards the pubes, and passes between the urethra and bladder above, and the rectum behind; at its internal extremity it surrounds the neck of the uterus, passing farther on its posterior than anterior surface. The structure of the vagina is composed exteriorly of an erectile vascular tissue lined by mucous membrane: this is irregular, and presents near the orifice the hymen, a crescentic fold, the remains of which form the *curunculæ myrtiformes;* this membrane possesses a number of mucous glands and follicles.

The *internal* organs of generation in the female are the *Uterus, Ovaries,* and *Fallopian tubes.*

The *Uterus* is situated in the pelvis, between the bladder and rectum; pyriform in shape, its superior fundus larger, is turned upwards and forwards, its inferior fundus directed a little backwards is rounded, and terminates by a slight expansion, behind which is the cervix uteri; the anterior and posterior surfaces somewhat flattened, are partially covered by peritoneum; its edges rounded afford attachment to the broad ligaments, and at the superior extremity to the Fallopian tubes and ligaments of the ovary.

The *structure* of the uterus is composed of an external serous partial covering from the peritoneum, beneath which is a peculiar firm tissue, about half an inch in depth, dense as cartilage, and composed chiefly of muscular fibres; its interior is lined by mucous membrane. The cavity of the uterus commences at the *os* or *mouth;* this is a transverse aperture, hence *os tincæ,* which leads into the general cavity; this widens a little in the centre, and branches off on either side, and superiorly into a small canal which leads into the Fallopian tubes.

The *Fallopian tubes,* one on each side, proceed from the superior and lateral portion of the uterus outwards to the extent of about four inches, contained in the broad ligament, and terminate in an irregular expansion, the *corpus fimbriatum;* the cavity of the Fallopian tube is small, and terminates on the corpus fimbriatum by a small aperture, *morsus diaboli.*

The *Ovaries*, or female testes, are attached also to the superior end of the uterus immediately behind the Fallopian tubes. They are contained in the broad ligament, and are attached to the uterus by an impervious ligament. The ovaries are covered by peritoneum, beneath which is a fibrous coat; the interior is composed of a number of vesicles; *Graafian vesicles*, from six to ten or twelve on each side.

The Uterus is supplied with blood by the uterine arts. from the int. iliac; its veins terminate in the int. iliac veins; its nerves are derived from the sympathetic. The ovaries receive their supply of blood from the spermatic arteries.

The *Ligamentum teres* is an analogue to the spermatic cord in the male, although it does not perform any special function. It is composed of cellular tissue, nerves, &c., and stretches from the upper ext. of the uterus, in front of the Fallopian tubes, to the labium pudendi.

The Perinæum in the female is comparatively unimportant; it extends from the fourchette backwards to the anus, and has muscles analogous to those in the male.

CHAPTER V.

THE INFERIOR OR LOWER EXTREMITIES.

are connected to the trunk by numerous muscles and ligaments; the former extend from the pelvis and spine, the latter from the ligaments of the hip-joint.

Beneath the integuments covering the lower extremities is found a quantity of superficial fascia, or loose subcutaneous cellular tissue. It contains numerous superficial blood vessels, nerves and lymphatic glands, of these the largest is the *int. saphena* vein, which may be traced from the foot upwards to its termination in the femoral vein. The arteries are chiefly branches from the femoral, the nerves from the anterior crural and lumbar plexus.

Beneath this is a dense layer of fascia, the fascia lata, the strongest in the body, which not only invests the muscles, but sends septa in between them. (See Fasciæ.)

Muscles of the Lower Extremity.

Of the Hip-Joint, 7.

Gluteus Maximus, Or. 1, post. fifth of the crista ilii, and rough surface beneath it, down to the super. semicircular ridge; 2, from the lumbar fascia and sacro-iliac ligaments; 3, from the sides of the sacrum and coccyx and post. sacro-iliac ligament; 4, from the fascia of the gluteus medius. *Ins.* by a flat thick tendon into an irregular longitudinal surface, leading from the great trochanter to the linea aspera of the femur, and into the fascia lata just below. *Use,* to extend, abduct, and rotate the thigh outwards; to make tense the fascia lata, and to fix the pelvis on the lower extremity.

ADDUCTORS.

Gluteus medius, Or. 1, from the ant. three-fourths of the christi ilii, and ant. sup. spinous process; 2, from the surface of the ilium between its two curved lines. *Ins.* into the outer part of the great trochanter. *Use,* to abduct the thigh; the posterior fibres rotate the limb outwards, the anterior inwards.

Gluteus minimus, Or. from the inf. curved line and surface of the ilium down to the sup. margin of the acetabulum; *Ins.* ant. half of the margin of the great trochanter. *Use,* similar to the last.

Pyriformis, Or. 1, by three slips from the ant. surfaces of the second, third, and fourth divisions of the sacrum; 2, from the ant, surface of the great sciatic ligament, and upper edge of the notch, *Ins.* post part of the margin of the great trochanter. *Use,* to abduct and rotate the thigh outwards.

Obturator internus, Or. 1, from the inner surface of the obturator fascia and fibrous arch of the obt. vessels; 2, from the bony margin of the foramen and surface of the ischium, between it and the sciatic notch; 3, from the upper edge of the true pelvis. *Ins.* into the digital fossa. *Use,* to abduct and rotate the thigh outwards.

Gemelli, (accessory fibres of the obt. int.) *Or,* the upper one, from the spine of the ischium; the lower one, from its tuberosity; they enclose the obt. tendon, into which, and into the digital fossa, they are inserted; the upper one is sometimes absent. *Use,* to abduct and rotate the thigh outwards,

Obturator externus, Or. ant. and lower part of the obturator membrane, and corresponding portions of the margin of the foramen; its tendon passes in a groove between the tuber ischii and the edge of the acetabulum, winds around the neck of the femur, to be *Ins.* into the digital fossa below the other muscles. *Use,* to abduct and rotate the thigh outwards.

The abductor and external rotator muscles of the thigh are 7 in number, and arise successively in a line which may be traced along the sides of the lumbar vert. crista ilii, pectineal line, body and ramus of the pubis, and remus and tuberosity of the ischium.

Psoas minor, Or. sides of the bodies of the last dorsal and the first (and sometimes the second) lumbar vertebræ: its narrow tendon expands to be *Ins.* into the ilio-pectineal line, and, by its outer edge, into the fascia iliaca: it tenses that fascia, and, by thus confining the two following muscles in their places, may be deemed their accessory. It also acts upon the vert. column and pelvis.

Psoas magnus, Or. by slips, the intervals between which give passage to the spinal nerves, from the sides of the bodies and intervert. substances of the last dorsal and the five lumbar vert., and from the roots of the corresponding transverse processes; *Ins* by a tendon equally shared by the iliacus, into the smaller trochanter. The lumbar plexus of nerves is imbedded in its substance. *Use*, to flex the thigh and rotate it outwards, to bend the body forwards.

Iliacus, Or. 1, from the margin and the entire surface of the iliac fossa; 2, from the two ant. spinous processes: 3, from the base of the sacrum; 4, from the capsule of the hip-joint. *Ins.* with the tendon of the psoas mag. into the smaller trochanter. *Use*, to flex and rotate the thigh upwards. These two muscles pass in front of the capsular lig. of the hip-joint, from which they are separated by a large bursa mucosa.

Pectineus, Or. 1, from the spine of the pubis and the pectineal line; 2, from the inf. surface of an aponeurosis continued from Gimbernaut's ligament. *Ins.* into the commencement of a line leading from the trochanter minor to the linea aspera; separated from the psoas by the femoral artery, as it lies in front of the quadratus femoris. *Use*, to adduct, flex, and rotate the thigh outwards.

Adductor longus. Apparently continuous at its origin with the pectineus, *Or.* by a flat tendon from the spine of the pubis; *Ins.* into the middle third of the linea aspera, in front of the abductor magnus.

Abductor Brevis, Or. body of the pubis below the spine and between the gracilis and obturator externus; *Ins.* in front of the abductor magnus, into the line leading from the trochanter minor to the linea aspera, below the *Ins.* of the pectineus. The three last muscles lie on a plane anterior to that occupied by the two following,

Adductor magnus, Or. 1, ramus of the pupis and ischium; 2, lower part of the tuberosity of the ischium. *Ins.* fleshy into the whole length of the middle ridge of the linea aspera, and into a line leading from it to the great trochanter above, and to the inner condyle below ; 2, by a tendon into a *tubercle* on the post. and upper part of the inner condyle. An opening exists, between the two portions, at the lower third of the thigh, formed especially for the transmission of the femoral vessels, to the sheath of which its edges are united. *Use,* the three adductors adduct, flex and rotate the thigh outwards.

Quadratus femoris, quadrilateral, *Or.* from the outer edge of the tuberosity of the ischium, and, frequently, from the ramus of that bone. *Ins.* into the line leading from the trochanter major to the linea aspera : it appears to be a continuation of the ad. mag., from which it is only separated by the int. circumflex artery. *Use,* to adduct and rotate the thigh outwards.

Muscles of the front and sides of the thigh, No. 7.

Tensor vaginæ femoris, Or. by a tendon, from the crista and ant. sup. spinous process of the ilium between the sartorius and glutæus medius; it is also attached to the aponeurosis of the latter muscle; it descends slightly backwards to the *Ins.* between two lamellæ of the fascia lata, at about the upper third of the thigh. *Use,* to make tense the fascia, and rotate the thigh inwards.

Sartorius, Or. ant. and sup. iliac spinous process, with the tensor fascia, and from half of the interspinous notch. Its tendon expands to be *Ins.* into the crista of the tibia, just below its tuberosity; its expansion covers the tendons of the gracilis and semi-tendinosus; it also strengthens the fascia of the leg. *Use,* to flex the thigh, adduct and rotate it inwards; also to flex the leg on the thigh obliquely so as to cross over the opposite limb.

Gracilis, Or. by a thin tendon, from the inner part of the pubis symphysis, from the spine, to the ramus of the ischium, internal to the abductor magnus. *Ins.* into the crista of the tibia, above the semi-tendinosus. The spreading of these muscles at their insertions is termed by the French anatomists, *patte d'Oie*. *Use,* to adduct the leg and thigh, to bend the knee, and turn the leg inwards.

Quadriceps extensor.

Rectus femoris, Or. 1, by a *straight tendon* from the inf. iliac spinous process; 2, by a *reflected tendon* from a fossa, just above the edge of the acetabulum. *Ins.* by a strong tendon which it shares with the vasti and crureus, into the ant. surface, and edge of the upper half of the patella. *Use,* to extend the leg on the thigh, to flex this on the pelvis.

Vastus externus, Or. 1, base and ant. part of the great trochanter; 2, from the tendinous ins, of the gluteus max. and short head of the biceps, and from the outer edge of the whole length of the linea aspera. *Ins.* into the external side of the common tendon and the patella: it covers the ext., part of the crureus at its origin. *Use,* to extend the knee, and rotate the leg outwards.

Vastus internus, Or,1, inter-trochanteric line, and whole length of the inner edge of the linea aspera, 2, inner surface of the femur. *Ins.* into the inner edge of the common tendon, and the patella, and by an aponeurosis, which covers the inner side of the knee-joint into the head of the tibia. *Use*, to extend the knee and rotate the leg inwards.

Crureus, inseparable from the last m., *Or.* fleshy from the ant. and ext. surface of the upper three-fourths of the femur, commencing at the ant. intertrochanteric line. *Ins.* upper edge of the patella with the common tendon, and into the synovial membrane behind it. A large bursa, which sometimes communicates with the joint, is placed between its tendon and the femur. The common tendon before it reaches the patella, is separable into three portions: the anterior one belongs to the rectus, the middle to the vastus externus, and the last to the vastus internus and crureus. The ligamentum patella continues them from the anterior surface and margin of the lower half of the patella, to the tuberosity of the tibia. *Use*, to extend the knee.

Muscles on the posterior part of the Thigh, 3 Flexors.

Biceps, flexor cruris, Or. 1, by its *long head*, in common with the semitendinosus, from the outer and posterior margins of the tuber ischii, immediately below the inferior gemellus; 2, by its *short head* from the linea aspera, from the ins. of the gluteus maximus, to within two inches of the outer condyle. *Ins.* head of the fibula. Its tendon also expands, to strengthen the ext. lat. lig. and fascia of the leg. *Use*, to flex the knee, the long head will extend the thigh and rotate the limb outwards.

Semitendinosus, Or. tuber ischii, with the long head of the biceps. *Ins.* into the crest of the tibia, beneath the expanded tendon of the sartorius, and below the tendon of the gracilis, to which it is joined. *Use*, to flex the knee and rotate the leg inwards; to extend the thigh.

Semimembranosus, Or. outer part of the tuber ischii, in front of the biceps and semitendinosus. *Ins.* back part of the inner tuberosity of the tibia: from this point a process is sent diagonally across the joint, to the inner part of the ext. condyle, so as to form the greater part of the lig. posticum; another winds, in a groove lined by synovial membrane, horizontally round the inner tuberosity of the tibia, into which it is inserted; and a third expands over the popliteus m. *Use*, to extend the thigh, to flex the knee, and rotate the leg inwards.

In the dissection of the anterior part of the thigh, the great vessels and nerves, viz., the femoral art., vein, and anterior crural nerve are found.

The *Femoral Art.* is the continuation of the ext. iliac; it passes beneath Poupart's ligament, having the vein to its inner side, the ant. crural nerve to its outer side and at some distance from it; from this the femoral art. passes downwards, backwards, and inwards, and finally escapes beneath the adductor magnus tendon into the popliteal space, and assumes the name of popliteal art.; its principal branch is the profunda; the femoral vein accompanies the art., at first lying internal to it, and then getting posterior to it. The anterior crural nerve descends into the thigh between the psoas magnus and iliacus internus muscles, and soon divides into a lash of filaments to supply the principal muscles on the anterior and lateral surfaces of the thigh; its most important branch is the int. saphenous nerve, which accompanies the femoral art. in its lower third, and leaves it to descend with the anastomotica magna to the knee, below which it joins the int. saphena vein.

The arteries met with on the back part of the hip and thigh are the gluteal, sciatic and pudic branches of the internal iliac; the first two descends into the thigh, the last is lost about the perinæum and organs of generation.

The nerves are the great and lesser sciatic nerves, branches of the sacarl plexus.

Section I.

Of the Leg.

Beneath the integuments of the leg, as in most other parts of the body, is found the superficial fascia; in it are contained, on the inner side the int. saphena vein and saphenous nerve; on the posterior surface the ext. saphena vein and post. saphenous nerve, will be found, but beneath the fascia of the leg at its upper part.

Beneath the superficial fascia is a strong aponeurosis, the fascia of the leg, attached on either side to the spines of the tibia and fibula. On raising this will be found the

Muscles of the Leg, 13. *Those in Front*, 3

Tibialis anticus, *Or.* 1. from the upper two-thirds or the ext. surface and crista of the tibia, as far as the outer tuberosity; 2, from the inner half of the interosseous ligament, from the fascia of the leg and septa. *Ins.* into a tubercle on the first *cuneiform* bone, and base of the first metatarsal bone. *Use*, to flex the ankle, adduct the foot, and to raise its inner edge from the ground.

Extensor communis digitorum, *Or*, 1, from the outer tuber tibiæ, with the tibialis anticus; 2, from the whole length of the surface of the fibula, in front of the interosseous ligament, and slightly from that lig., from the crural fascia and septa. *Ins.* by four tendons, into the phalanges of the four lesser toes, like the corresponding m. of the arm. The outer fibres of the inf. third of the muscle are attached to a fifth tendon. *Use*, to extend the toes and flex the ankle.

Extensor proprius pollicis, *Or.* middle third of the ant. surface of the fibula covered by the last m., slightly from the interosseous ligament, from the crural fascia and septa; *Ins.* by two fasciculi into the first and second phalanges. *Use*, to extend the great toe and flex the ankle.

Muscles on the outer part of the Leg, 3.

Peroneus longus, *Or.* 1, outer and fore part of the head of the fibula, and ext. surface of the upper two-thirds of that bone; 2, slightly, from the inner tuber tibiæ, with the common extensor from the crural fascia and septa. *Ins.* into a tubercle on the outer part of the base of the first metatarsal bone; its tendon crosses the foot above all the other soft parts, lodged in a groove in the cuboid bone. *Use*, to extend the ankle, turn the foot outwards, and raise its outer edge.

Peroneus brevis, *Or.* lower half of the outer and posterior surface of the fibula; *Ins.* base of the fifth (sometimes also of the fourth) metatarsal bone. *Use*, similar to the last.

Peroneus tertius. The external fibres of the lower third of the ext. com. m. terminate in a tendon which is *ins.* into the upper surface of the fifth metatarsal bone, just in front of the last muscle. *Use*, to flex the ankle, raise its outer edge, and abduct the foot.

Muscles on the back of the Leg, 7.
Superficial set, 4.

Gastrocnemius, *Or.* by two heads, each of which is attached to a depression on the upper, outer, and back part of each condyle, and condyloid ridges; the inner one, the larger, just behind the tendon of the abductor mag.; the outer one, just above the tendon of the popliteus. The resulting two fleshy bellies are united in a raphè, a little below the joint, and terminate at the middle of the leg, in a flat tendon, which almost immediately becomes a part of the tend. Achillis. *Use*, to extend the ankle-joint, to raise the body in walking, to flex the knee-joint.

Plantaris, Or. ext. condyloid line, and fibrous capsule over the outer condyle; its long narrow tendon runs to the inner edge of the tendo Achillis, to be *Ins.* into the calcaneum, or merely into the fatty tissue, before the insertion of that tendon. Sometimes double. *Use,* to aid the preceding.

Soleus, Or. 1, *tendinous,* from the inner and back part of the head of the tibia, and *aponeurotic,* from the post. surface and outer edge of the upper third, of the fibula; 2, from the popliteus fascia post. oblique line of the tibia, and from a third of its inner edge: just below, its tendon, with that of the gastrocnemius, forms the tendo Achillis, which is *Ins.* into the lower, and back part of the calcaneum. *Use,* to extend the ankle and assist the progression by raising the os calcis from the ground.

Popliteus, Or. narrow, from a fossa on the back part of the outer condyle of the femur; *Ins.* broad, into all the triangular surface of the tibia, above its oblique line: the tendon is covered by the ext. lat. ligament, and surrounded by the synovial membrane of the joint. *Use,* to flex the knee, and when flexed to turn the leg inwards.

Deep set, 3.

Tibialis posticus, lies in the posterior interosseous space; *Or.* by two fleshy slips, between which passes the ant. tib. art.; one from the oblique line of the tibia, below the popliteus, soleus, and common flexor; the other from the inner edge of the fibula, below the soleus; also, the whole of the interosseous ligament, and adjoining surface of the fibula. *Ins.* with a sesamoid bone into a tubercle on the inner side of the *scaphoid* bone. *Use,* to extend the ankle, and raise the inner edge of the foot from the ground.

Flexor digitorum communis perforans, Or. posterior surface of the tibia, between the oblique line and a point two inches above the inner malleolus; *Ins.* into the last phalanges of the four smaller toes, like the corresponding muscles of the fingers. *Use*, to flex the toes, to extend the ankle, and assist in raising the body from the ground in progression.

Flexor pollicis pedis, the largest muscle of this set, *Or.* inf. two-thirds of the post. surface of the *fibula: Ins.* into the last phalanx of the great toe, like the corresponding muscle of the thumb. *Use*, to flex the great toe and extend the ankle.

Muscles of the Foot.
Of the Dorsum, 1.

Dorsalis pedis, Or. from the outer and upper part of the calcaneum, and adjoining portion of the ext. calcaneo-cuboid ligament; *Ins.* by four tendons, of which the three last are *ins.* into the expansion of the common extensor tendons on the first phalanges of the second, third, and fourth toes, and the first into the base of the first phalanx of the great toe. *Use*, to extend the four inner toes.

The Sole of the Foot.

The integuments covering the sole of the foot are remarkably dense, particularly where exposed to pressure, as on the heel and outer side of the foot; beneath these is a thick layer of adipose substance, collected into small masses so as to form an elastic cushion to preserve the subjacent parts from injury in progression; beneath this is a dense layer of fibrous tissue, the *plantar aponeurosis;* triangular in shape it is attached posteriorly by its apex to the tubercles on the os calcis, passes forwards, spreads out, and divides into three portions; of these the central is the strongest, the outer the next in strength, and the inner the weakest; these send in processes between the muscles; the lateral are attached to the sides of the tarsus and metatarsus, the central portion divides anteriorly into five fasciculi, which again subdivide, and are attached to the lateral ligaments of the metatarso-phalangeal articulations, leaving intervals for the passage of the digital tendons, vessels, and nerves. Beneath this are the

Planter Muscles, 19.

Of the Inner Edge, 4.

Abductor pollicis, Or. 1, from the *inner* tubercle of the calcaneum; 2, from the int. annular lig. and plantar fascia. *Ins.* into the int. sesam. bone of the base of the first phalanx of the great toe, with the inner head of the flexor brevis. *Use,* to abduct the great toe.

Flexor brevis pollicis, Or. from the two last cuneiform bones, and from the ant. part of the calcaneum and their ligaments; *Ins.* by two heads with the abductor and adductor pollicis, into the sesamoid bones and base of the first phalanx. *Use,* to flex the first joint of the great toe.

Adductor pollicis is placed in the hollow formed by the lower surfaces of the four last metatarsal bones; *Or.* cuboid and *bases* of the three last metatarsal bones, and from the sheath of the peroneus long. tendon; *Ins.* with the outer head of the last m. into the ext. sesamoid bone. *Use,* to abduct and flex the great toe.

Adductor secundus vel transversus pedis, Or. from the *head* of the last metatarsal bone, and phalangeal ligaments of the two next; *Ins.* outer edge of the first phalanx of the great toe, close to the last muscle.

Muscles of the outer Edge of the Foot, 2–3.

Abductor minimi digiti, Or. from the ext. tubercle of the calcaneum, and sometimes from the base of the last metatarsal bone; *Ins.* outer side of the base of the first phalanx of the little toe. *Use,* to abduct and flex the little toe.

Flexor digiti minimi resembles an interosseous m. *Or.* base of the fifth metatarsal bone and adjoining ligaments; *Ins.* on the outer part of the base of the first phalanx. A few fasciculi from the same origins may sometimes be traced to the whole of the outer edge of the fifth metatarsal bone. These might be called *opponens digiti minimi*. *Use*, to flex and adduct the little toe.

Muscles in the Middle of the Foot, 13.

Flexor brevis perforatus, Or. 1, from the calcaneum, between the tubercles; 2, from the plantar fascia, and septum on each side. *Ins.* by slips, into the four smaller toes, like the flexor perforatus of the hand. *Use*, to flex the four outer toes.

Flexor accessorius, Or. inferior surface of the calcan., and calcan. scaphoid ligament; *Ins.* 1, into the outer edge of the tendon of the common flexor; 2, strengthened by a process from the tendon of flex. long. pol., into the upper surfaces of each tendon of the common flexor. *Use*, to flex the toes with or without the long flexor.

Lumbricales resemble precisely the corresponding fasciculi in the hand. *Use*, to adduct and flex the four outer toes.

The Interossei are seven in number, as in the hand; viz., four dorsal and three plantar, the former being abductors, the latter adductors; but, in the foot, the median line, or axis of the second toe, must be taken as the fixed point. The first dorsal interosseous muscle is attached to the tibial side of the second toe; the second to the fibular side of the same toe; the third, to the fibular side of the third toe; the fourth, to the fibular side of the fourth toe. The plantar are attached, the first, to the tibial side of the third toe; the second, to the tibial of the fourth toe; the third, to the tibial side of the fifth or little toe.

The relations of the parts about the ankle joint, are as follows: *A. Posteriorly*, 1, the inner malleolus, and the groove and sheath containing the tendons of the tibialis posticus and common flexor; 2, the tendon of the flexor pollicis, which lies close to the calcaneum, separated by a space of an inch and a quarter from the last-named tendons; 3, the post. tibial vessels and nerve in the middle of this space, or somewhat nearer the flex. pol. tendon, and superficially, the intern. saphen. nerve and vein; 4, the tendon of the plantaris, tendo Achillis, and tuberosity of the calcaneum; 5, the communicans tibiæ, nerve, ext. saphena vein, and more deeply, the termination of the peroneal artery; 6, the ext. malleolus, and immediately behind it, the tendons of peronei long et brevis, in a single groove and sheath. *B. Anteriorly*, 1, the tendon of the tibialis anticus; 2, that of the ext. pollicis; 3, surfaces of the tibia and astragalus supporting the ant. tibial artery and nerve; 4, tendons of the common extensor, and tendon of the peroneus tertius.

In the dissection of the leg and foot, the principal vessels and nerves met with are the anterior and posterior tibial and their several branches.

The femoral art. having entered the popliteal space assumes the name of popliteal; this descends nearly in the centre of the space, and at the lower margin of the popliteus muscle divides into its two terminating branches, the anterior tibial, the smaller, and the posterior tibial. The *anterior tibial* art. passes forwards through the interosseous space and descends on the anterior surface of the interosseous membrane, passes beneath the anterior annular ligament of the ankle-joint, runs forwards and inwards to the cleft between the first and second metatarsal bones, and sinks into the sole of the foot to anastomose with the plantar arteries; it is accompanied in the lower two-thirds of its course by the ant. tibial nerve, a branch from the peroneal or fibular.

The *posterior tibial art.* descends obliquely inwards along the posterior surface of the leg, resting on the deep muscles, passes behind the internal malleolus, enters the sole of the foot, and terminates in the ext. and int. plantar arteries, of these the external is the larger and supplies three and a half toes from the little toe inwards, the int. plantar art. supplies the one and a half inner toes.

The posterior tib. art. is accompanied by the post. tib. nerve, which terminates like it in the side of the foot in two plantar nerves; of these the internal is the larger, and supplies the three and a half inner toes, the ext. supplies the remaining one and a half toes.

Shortly after its origin, the post. tib. art. gives off its largest branch, the *fibular;* this descends outwards along the back of the fibula, and terminates about the ankle-joint. It is not accompanied by any nerve.

CHAPTER VI.

FASCIÆ.

Superficial fascia. An areolar tissue, connected by numerous filaments to the dermis; its principal use is to render the skin movable on the subjacent parts, and to protect the subcutaneous vessels and nerves; it may also be considered as the connecting medium between the skin and aponeurosis. The subcutaneous cellular tissue, superficial fascia, and the proper investment of muscles, tendons, &c., are modifications of the same tissue, and in some regions it is difficult to tell where the one begins or the other ends; in others the investing fasciæ are sufficiently distinct, but no line of demarkation distinguishes the subcutaneous tissue from the so-called superficial fascia, the separation is always more or less arbitrary: thus, in the inguinal region, where it forms a layer of some importance, the adipose substance and cellules, so frequently developed in large quantities beneath the skin, must always be cut through, in raising the latter membrane, in order to demonstrate a superficial fascia.

Fascia, aponeurosis. These terms are almost indiscriminately employed to designate those layers of condensed cellular membranes, which, as sheaths, septa, or linings, preserve the various organs of the body in their forms and relations. The *fascia transversalis* is an example of the latter, and the fascia lata is the best specimen of a sheath. This fascia surrounds the whole inferior extremity, and sends numerous processes or septa from its inferior surface, some of which form secondary sheaths for the muscles, vessels, and nerves of the limb. If all these organs could be removed without injury to an aponeurotic investment of this kind, a faithful outline of their form and relative position would still remain; in fact, the whole body, if it could be subjected to such a process, would represent a skeleton of cavities and tubes, admirably adapted for the protection and support of their respective organs.

Fascia cervicalis (superficialis.) The superficial fascia of this region contains a cutaneous muscle, the *platysma myoides;* it may be traced from the median line completely round the neck; above, it is lost with the platysma on the face; and below, on the muscles of the chest; on the sterno-mastoid muscle it is identified with the next fascia; both are intimately united elsewhere, but can be separated without much difficulty.

Fascia cerv. profunda. A dense aponeurotic line extends from the symphysis of the chin to the os hyoides, and onwards to the inner part of the first bone of the sternum: from each side of this line the fascia may be traced round the neck to its posterior part. It furnishes a sheath for the sterno-mastoid muscle, and another for the submaxiliary gland, which is thus separated from the parotid; its processes form a sheath for the carotid artery, jugular vein, and vagus nerve, another for the omo-hyoid muscle, by which it is enabled to preserve its angular directions, and several less remarkable for the remaining organs, lymphatics, &c., in this region. A layer separates the sternal muscles from the trachea, and descends to be connected with a dense tissue, which so strongly ties the great vessels to the inside of the sternum; according to Todd and Bowman, it even reaches the pericardium: above, the deep cervical fascia is attached to the base of the jaw, and the stylo-maxiliary ligament, and over the parotid to the zygoma; below, to the inner side of the clavicle and first bone of the sternum.

Abdominal Fasciæ and Aponeurosis.

The *superficial fascia* deserves most attention in the inguinal and crural regions; there it presents a well-developed layer, which is strongly adherent a little below Poupart's ligament; it descends on the thigh, and, on the cord and testicle, to form one of the scrotal coverings. In early life it is said to occupy the inguinal canal, as a gubernaculum for the descent of the testis. Its crural portion is loaded with lymphatic glands and small vessels.

Sheath of the Rectus. The int. obliquus muscle terminates in an aponeurosis, which at the *linea semilunaris* is intimately united to that of the transversalis m., and then splits into two layers; one, closely united to the aponeurosis of the ext. oblique, passes to the linea alba in *front* of the rectus; the other, with the aponeurosis of the transversalis, to the same line *behind* that muscle: but where the rectus covers the ribs, and from a point midway between the umbilicus and pubes to the pubes, the sheath wants its post. paries, inasmuch as the four conjoined aponeuroses there pass in front of the muscle.

Sheath of the quadratus lumb., lumbar fasciæ. At the outer edge of the quadratus m., the inter. oblique and transversalis also terminate in one aponeurosis, which immediately afterwards splits into three layers. The *posterior one* passes behind the common origin of the longissimus dorsi m. to the tips of the lumbar spinous processes: it is confounded with the apon. of the ser. post. inf. and latissimus dorsi muscles. The *middle one* passes between the long. dorsi m. and the quadratus lumbar., and is inserted into the tips of the lumbar transverse processes. The *internal*, thin and weak, better understood as a continuation of the *fascia transversalis*, which it so much resembles, passes in front of the quadratus to the transverse processes, and then over the psoæ to the bodies of the lumb. vertebræ.

Ligamentum arcuatum (*Pseud.*) The upper part of the *fascia transv. post.* arching over the quadratus muscle, from the tip of the last rib to the transverse process of the first lumbar vertebra.

Lig. arcuat (*verum*,) the upper part of the same fascia arching over the psoæ, from the transverse process of the first lumbar vertebra to the body of the second. Both these ligaments afford attachments to the diaphragm, on the under surface of which the fascia trans. post. is continued.

Fascia transversalis (*anterior*.) Hardly perceptible at the umbilicus, it increases in density as it descends, and at the inguinal region appears as a dense shining membrane intimately adherent to the fibres of the muscle (it is a muscular fascia.) Behind the pubis, and, at Poupart's ligament, external to the femoral vessels, it joins the pelvic fascia; where it meets the vessels, it joins the superficial division of the fascia lata (*fascia cribrif.*) to form the ant. paries of the crural canal. The spermatic cord derives an investment (*fascia prop.*) from it, in its passage through the inguinal canal.

Pelvic fascia. A fine dense muscular fascia, resembling the fascia transversalis, to which it is connected along the base of the sacrum, crista ilii, Poupart's ligament external to the vessels, and at the pubis. It covers the iliacus muscle, as the *iliacus fascia*, and passes over the psoæ, *beneath* the iliac vessels, sending a thin layer in front of these, to descend some way on the obturator m. as the *internal obturator fascia*, but, at a line extending from the pubis to the spine of the ischium, it is reflected upon the upper surface of the levator ani m., which guides it to the sides of the bladder forming the true lateral ligaments of this viscus, vagina, and rectum: behind, it is reflected from the sacrum and coccygeus m. to the rectum; and before, from the body of the pubis and Wilson's muscles, as the *true vesicle ligaments*, to the neck of the bladder and prostate gland. This portion might be called the *inner*, or deep *perineal fascia*.

Perineal fasciæ, 1, the *superficial*, is analogous to the superficial fascia in other parts. 2, the *perineal*.

The fascia is continuous in front with that of the scrotum: on each side it is firmly adherent to the rami of the ischium and pubis; behind, at a line extending from one tuber ischii to the other, it is also adherent to the middle perineal fascia; here it is said to terminate (Blandin,) but it may still be traced, covering the levator ani, and filling up with a mass of fat, the triangular space on the inside of the tuber ischii, formed by that muscle and the lower part of the obturator internus. The superficial sphincter ani is developed in it.

The posteriorly *deep fascia* of the perineum, improperly called *triangular* ligament of the urethra, arises from the sub-pubic ligament, rami of the ischium and pubis, tuber ischii, inner edge of the great sacrosciatic ligament, and coccyx: from these attachments it passes inwards to the median line. That part which is *before* the transverse line is composed of strong transverse fibres; to this the term *triangular* ligament is applied. Behind this line, it first ascends on the inner side of the obturator internus as far as the line of reflection of the obturator portion of the pelvic fascia, and then turns down on the lower surface of the levator ani m. The triangular space lies between these two portions of fascia. The middle part of the levator ani arises from the line, where the obturator fascia from the pelvic, and that from the middle perineal fascia, are united. The ejaculatores semin., erectores penis, and transversales m., superficial perineal nerves and vessels, are enclosed between the perineal fascia and triangular ligament, and the levator ani, coccygei, and Wilson's muscles, between the latter and pelvic fascia. The middle perineal fascia bifurcates at the rami ischii et pubis, to enclose the internal pubic artery. Fluid effused in the anterior perineal sheath *before* the transverse line, will infiltrate the cellular tissue of the scrotum.

Fascia lata. A dense, fibrous sheath, enclosing the lower extremity. Numerous septa and sheaths are sent from this fascia to the vessels and muscles of the limb. The most remarkable are—1. The *Ext. intermuscular septum*, which extends from the trochanter major to the ext. condyle; its inner edge is attached to the whole length of the linea aspera, and separates the vastus ext. from the short head of the biceps, to both of which it gives attachment. 2. The *Int. intermuscular septum* extends from the inter-trochanteric line to the internal condyle. It separates the adductors from the vastus int. The femoral vessels and saphenous nerve are enclosed in a sheath derived from the fascia lata, the upper part of which is the *crural canal.* The lower part of the sheath is formed by a septum, which is sent down from the fascia to bifurcate and enclose the vessels; but, above, the fascia (supposing it to pass inwards) comes in contact with the vessels enclosed in their proper sheath, and at a falciform line (which is produced by the tension of the fascia between Gimbernaut's ligament, Poupart's ligament, and external side of the vessels,) sends off a fascia (*cribriform*) in front of the vessels and then reflects itself outwards and downwards to pass under them, in order to join the fascia cribiformis at their inner side; thus reunited, the fascia passses on to the inner side of the thigh. It now appears that the *crural canal* is formed by an arrangement of the fascia lata, precisely similar to what takes place in forming any other of its simple sheaths. The fascia transversalis joins the iliac portion of the pelvic fascia; 1st, along the crista ilii and Poupart's ligament, as far as the outer side of the femoral vessels; *i. e.*, a point midway betweeen the ant. sup. spin. process of the ilium and pubic symphysis; and 2d, from the outer edge of Gimbernaut's ligament to the pubis; a triangular opening is thus left, its base corresponding to the outer or free edge of Gimbernaut's ligament, its apex to the middle point above mentioned; its anterior side, being the inner half of Poupart's ligament, its posterior, the attachment of the pectineus muscle. The anterior and upper part of the femoral sheath, or the cribriform fascia, is connected by means of the falciform fold to the fascia transversalis at the anterior transverse half of the triangle, and the posterior part of the sheath to the fascia pelvica, at the posterior transverse half; the femoral vessels, enclosed in a proper sheath, occupy the outer part of this canal, leaving between them and the free edge of Gimbernaut's ligament a space through which the intestine makes its way. This space only extends to the point where the saphena vein passes through the cribriform fascia to join the femoral vein, and, except at the inner side of the vessels, the *ext. sheath* firmly adheres to the proper sheath. It follows, 1st, that this space is the real crural canal; 2d, the sac cannot descend lower than the termination of the saphena vein; 3d, the upper portion of the cribriform fascia, being the most yielding, is pushed before the sac, which will, in consequence, turn upwards over the edge of the falx; 4th, as the falx is connected to Gimbernaut's ligament, the division of one will tend to relax both.

The fascia lata gives a sheath to the sartorius muscle, another common to the three hamstring muscles several to the abductor muscles, a very strong one to the tensor femorism. Superiorly, the anterior portion of the fascia lata is united to Poupart's ligament, &c., as described; posteriorly, it adheres to the crista ilii, and over the sacrum to the posterior lumbar fascia. Below, the fascia is continued over the knee-joint, forming a fibrous capsule, adherent to the patella and its ligament, the femoral condyles, and muscular tendons. It is strengthened on the inner side of the joint by an aponeurosis, derived from the vastus internus.

Crural fascia, continuous with the fascia lata. This fascia surrounds the leg; it is particularly strong over the extensor muscle, to which it affords some extent of attachment: a strong septum separates the extensors from the peroneal muscles, another is placed between the latter and the posterior muscles. The fascia is thus subdivided into three principal sheaths; the posterior one is still further divided by a strong fascia, which passes transversely between the superficial and the deep set of muscles and posterior tibial vessels.

Annular ligaments. At the ankle-joint the fascia is strengthened by transverse fibres; one set the *ant. ann. lig.*, arises from the ext. part of the calcaneum by a small but dense extremity, and, expanding as it runs inwards, sends a branch to the inner malleolus, and another across the foot to the inner edge of the plantar fascia; the first gives a sheath to the tendon of the tibalist ant. and common extnsor and a partial one to the exten. pollices. The second branch forms a distinct sheath for each. A second set, *int. annular ligament*, arises from the internal mallelous, and terminates, broad, into the inner side of the calcaneum, and plantar fascia. The third set, *ext. annular ligament*, extends from the outer malleolus to the calcaneum. Distinct sheaths are formed for the tendons, vessels and nerves passing behind each malleolus.

The *dorsal aponeurosis* of the foot covers the dorsum, uniting with the plantar fascia at the edges of the foot, and terminating at the heads of the metatarsal bones.

The *plantar aponeurosis* is divisible into three portions; the *middle* commences at the inner tuberosity of the calcaneum, and proceeds, gradually becoming broader, to the heads of the metatarsal bones, where it divides into four branches, each of which nearly ensheaths the corresponding flexor tendon, and is inserted into the edges of the dorsal expansion of the first phalanges. This portion sends a septum on each side, which separates the muscles in the middle from those on either edge of the sole of the foot; the *inner portion* covers the inner muscles of the foot; it arises from the int. annular ligament; internally, it joins the dorsal aponeurosis, and externally, the internal septum; the *outer portion*, much stronger, covers the muscles on the outer edge of the foot; it arises from the calcaneum; internally it joins the ext. plantar septum, and externally the dorsal aponeurosis; it is also firmly attached to the base of the fifth metatarsal bone. Numerous septa from the upper surface of the plantar fascia pass between the plantar muscles and tendons to be attached to the tarsal and to the edges of the metatarsal bones.

The arm and shoulder also possess aponeurotic investments. The supra spinatus, intra spinatus, and subscapular fossæ are converted into osteo-fibrous cases by strong aponeuroses, from which the respective muscles derive attachments. The infra spinatus apon. bifurcates to enclose the deltoid; the superficial layer joins the brachial fascia, and the deep one is attached to the short head of the biceps.

The *brachial fascia* is firmly attached to the margins of the axilla, and appears also to be connected to the acromion and spine of the scapula. It encloses the arm, and is continuous below with the fascia of the fore-arm; the majority of its fibres are circular. Its most remarkable processes are, 1st, *int. intermuscular septum*, which is attached to the posterior lip of the bicipital groove below the teres major, and the inner ridge and condyle of the humerus The *external septum* commences at the anterior lip of the bicipital groove, where it is united to the tendon of the deltoid, and is firmly attached to the outer condyloid ridge and condyle. The muscles and vessels on the anterior and posterior part of the arm are thus enclosed by two large distinct sheaths, from which small ones are derived for each muscle or vessel. The brachial vessels and median nerve have a single sheath. The aponeurosis of the *fore-arm* is strengthened above by the expansion of the biceps tendon; a transverse septum separates the superficial from the deep set of muscles on the anterior part of the arm. A similar septum passes between the superficial and the deep-seated extensor muscles on the back of the arm.

The *dorsal annular ligament* of the wrist is composed of thin parallel fibres, which extend from the pisiform bone and palmar fascia to the outer edge of the radius. It forms six distinct sheaths: 1st, one for the tendons of the extens. os. metacarpi pollicis and prima internodii; 2, radial extensors; 3, exten. secundi internodii; 4, exten. communis et exten. indicis; 5, ext. min. digiti; 6, exten. carpi ulnaris.

The *ant. annular lig.* is a thick fibrous band which arises from the pisiform and unciform bones, and is inserted into the trapezium and scaphoid bones. Most of the muscles of the thumb and little finger arise from its anterior surface. It presents a partial sheath for the tendons of the flex. carp. rad., and only one for the tendons of the flexor muscles and median nerve.

The *palmar fascia* arises from the ant. annular lig. and tendon of the palmaris longus; it is composed of strong diverging fibres, which are disposed over the muscles and tendons in the middle and edges of the hand, precisely as the plantar fascia in the foot.

Composed of transverse fibres, there is a *dorsal fascia*, derived from the dorsal annular ligament; it covers the back of the hand. A loose but elastic cellular tissue serves the purpose of synovial sheaths for the extensor tendons.

CHAPTER VII.

THE VASCULAR SYSTEM.

UNDER this head are comprised the heart, arteries, and veins, the organs for the circulation of the blood; and the lymphatics, which absorb the lymph and chyle, and convey them into the sanguineous circulation.

SECTION I.

THE HEART

is of a flattened conical shape, and placed in an oblique direction between the two lungs, where it rests upon the central tendon of the diaphragm. Its base, turned upwards and backwards, lies close to the spine, and its apex nearly touches the fifth and sixth costal cartilages of the left side: its edges are turned to either side. The division of the heart into two auricles and two ventricles is indicated, 1st, by a circular furrow, which surrounds the heart near its base, and separates the auricular from the ventricular portions; and 2dly, by a vertical furrow, which, after dividing the former portion into two auricles, descends on the anterior and posterior surfaces of the latter portion, meeting at a little to the right of the apex. The anterior surface of the ventricular portion is unequally divided by the anterior portion of the ventrical furrow: the greater part, which is on the right of the furrow, belongs to the right ventricle, the smaller to the left. The inferior surface, which rests upon the diaphragm, is nearly equally divided by the posterior furrow. The right margin of the heart, which also rests on the diaphragm, thus forms a part of the right ventricle; and the left, thick and round, which is lodged in a hollow in the left lung, forms the greater part of the left.

The following parts present themselves from before backwards on the base of the heart; 1, the *pulmonary artery* and *infundibulum*, behind which, and at first, partly concealed by them; 2, the *arota;* 3, the circular furrow; 4, the two auricles, and their appendages, that of the right being in front, and the left behind.

The *right ventricle* (anterior and inferior) is triangular; its parietes, which are thin in comparison with those of the left, are remarkable internally for their reticulated appearance, which is produced by the intercrossing of the *carneæ columnæ;* one set of these, few in number, is attached to the parietes of the ventricle by one end only, and by the other to the chordæ tendineæ and auriculo ventricular valve; the second by both ends; the third, in their whole length, by their sides.

The *auricular orifice* posterior, and to the right side of the base, is elliptical, and provided with an auricular membranous fold (*tricuspid valve,*) which projects into the ventricle; the valve, although it has even more than three angles, is really mitral, each division corresponding to the ant. and post. half of the ellipsis.

The *chordæ tendineæ* arise from the ends of the first set of columns, and from the sides of the ventricle, particularly the septum cordis; as their office is to tense and close the valves, one set is inserted into the under surfaces of the valves, and other sets, coming from opposite sides of the ventricle, cross each other to be inserted into the edge of the opposite valve.

The *orifice of the pulmonary artery* is anterior, and to the left side of the base, and separated from the auricular orifice by a portion of the tricuspid valve: it is guarded by three folds (*sigmoid valves,*) each of which presents a small nodule (corpus aurantii) in its centre.

Left ventricle (posterior and superior) differs from the right; first, it exceeds the latter at the apex, but is embraced by it at the base; secondly, it is conical and uniformly convex; it holds somewhat *more* than the right (Cruveilhier,) somewhat *less* (Bouillaud;) thirdly, some of its columns are bifurcated, and they enter into more complex union with each other: its sides are thicker in the prop. of three to one; fourthly, the valve is more evidently mitral, and the right half alone separates the auricular from the *aortic orifice*; the latter has three valves, like those of the pulmonary, but the two orifices are so near each other that this half of the mitral valve and the adjoining aortic valve have a common base.

The *right auricle* has three sides, an anterior, posterior, and that formed by the septum auric. It has five openings. 1. the auriculo ventricular; 2, that of the *vena cava descendens*, which looks downwards and slightly backwards; 3, that of the *vena cava inf.*, which forms a right angle with the trunk to open into the auricle close to the septum aur., through the post. side. It is guarded by a valve (*Eustachian*,) which surrounds the anterior half, sometimes two-thirds, of the orifice; one extremity seems to be continuous with the *annulus ovalis*, the other is lost in the margin of the orifice; 4, that of the *coronary vein*; this is placed before the latter, and separated from it by the Eustachian valve; it is provided by a valve (*valvula Thesbesiana*; 5, that of the *foramen ovale*, or rather the *fossa ovalis*, bounded above and in front by a semicircular prominent margin, improperly called *annulus*. A projecting muscular fasciculus, said to be the seat of a tubercle (*tuber. Loweri,*) separates the openings of the two *cavæ* on the right side; another, in a vertical direction, divides the auricle into its sinus, which comprehends these openings, and its proper *auricular portion*. The auricular portion presents numerous vertical muscular fasciculi (*musculi pectinati,*) which are crossed by smaller oblique fibres, so as to give the surface a reticulated appearance. The foramina (*Thebesii*) seen in this auricle are small orifices of a few areolar spaces.

The *left auricle* differs from the right; 1, it is somewhat less capacious; 2, in shape it is an irregular cuboid; 3, the auriculo ventricular orifice is less in size; 4, it has four venous openings (pulmonary) without valves, two opening on the right side and two on the left; 5, it communicates only by a narrow orifice with its appendix; 6, it presents no vestige of fossa ovalis, unless, which is sometimes the case, the two auricles still communicate by a narrow oblique canal.

Structure of the heart. The basement or skeleton of the heart is formed, 1, by two fibrous zones, which constitute the auriculo ventricular orifices, and send expansions into the mitral and tricuspid valves; 2, by two arterial zones, which in the same way surround the arterial orifices, and send expansions into the sigmoid valves, and three thin, but strong processes, which fill up the intervals between the three festoons at the base of each artery, and strengthen their attachment to the ventricles. These two auriculo ventricular orifices are closely united in the same plane, and the aortic orifice, placed in their angle of separation in front, is confounded with them in its posterior half. This point is occupied by a cartilaginous or even osseous arch (os cordis) in some animals. The *pulmonary orifice* is anterior to, and five or six lines higher than the latter.

The *muscular* portion of the heart arises from these zones. The ventricular fibres are separable into three distinct layers. The *middle layer* forms a distinct conical tube, open at both ends, for each ventricle; the *ventricular septum* is formed by their adossement. The *outer layer*, common to both ventricles, descends in a spiral direction; those covering the right ventricle to the left, and those on the left ventricle to the right, as far as the apex. The two layers, twisting half round each other, now enter the openings at the apex of each tube formed by the middle layer, and divide into two sets to form the *inner layer* of each ventricle. The first set pass upwards, to line the *opposite* side of the ventricle, and the second, variously contorted, ascend on the *same* side; in both cases those fibres which were superficial in the outer layer become the deep-seated in the inner one. The *columnæ carneæ* seem to be made up of portions of the first, and of spiral fibres of the second set.

Auricular fibres. A *common* layer extends transversely across the anterior surfaces of both auricles. Of the *layer proper* to each auricle, a circular set of fibres is seen near the auriculo ventricular openings, and a second set, running an oblique course, is arranged in arches, which are fixed by their ends to the ventricular zones, their free portions surrounding the venous orifices, and projecting between them. The *annulus ovalis* is formed by some muscular arches of the septum. If the common ventricular fibres be cautiously cut through at the vertical furrow, the ventricular tubes may be separated from each other entire.

Development. At an early period of conception, the heart fills the thoracic cavity, and, up to three months, its direction is vertical, and its shape round and symmetrical. The ventricles are very minute, and the right auricle in size equals the rest of the organ; subsequently its four cavities are nearly equal. The auricular septum is wanting, and the ventricular is deficient at its upper part; the auricles afterwards communicate by the *foramen ovale*, and the ventricles indirectly by the *ductus arteriosus*. The Eustachian valve covers part of the orifice of the sup. cava, as well as that of the lower one (Martin St. Ange;) as development proceeds, the foramen ovale is gradually closed by a membrane, and the Eustachian valve diminishes in the same proportion.

The heart is enclosed in two distinct investments,

1. The *fibrous pericardium*. This membrane is formed of aponeurotic fibres; it is conical in shape, its base corresponding to the tendon of the diaphragm, to which it adheres in the adult; its apex surrounds the origins of the large vessels, upon which it is gradually lost. Its size is said to equal that of the heart in the greatest state of distention.

2. The *serous pericardium*. A complete sac of serous membrane which lines the former envelope, and is reflected from it on the great vessels and heart; it passes up on the former to some distance, especially on the aorta and pulmonary art., least on the inf. vena cava.

The *endocardium* (Bouillaud) is a white pellucid membrane, resembling the arachnoid; it lines the cavities of the heart, covers the valves and tendinous cords, and is continuous with the inner coat of the arteries. In the neighborhood of the zones and in some of the areolæ of the auricles, it comes into contact with the serous pericardium. The valves of the heart are formed by folds of this membrane, with a little fibrous tissue at their free margins. The aortic and pulmonary valves are named *semilunar* and present each, in the centre of its free margin, a slight eminence, the corpus aurantii.

The heart is supplied with blood by the *coronary arteries*. These arise from the aorta, immediately above its opening and behind the aortic valves; the *right* supplies the right auricle, the posterior part of the ventricles, and the thin edge of the heart; it sinks into the right auriculo ventricular groove, winds round to the back of the heart, sends a long branch along the anterior thin edge of the heart, runs along the groove on the posterior surface, and terminates by inosculating with the *left;* this supplies the left auricle, left ventricle, and septum. One branch runs in the left auriculo ventricular groove, the others along the anterior surface of the heart, opposite the septum and terminates near its apex by inosculating with the right. The coronary veins terminate by one large opening in the right auricle, between the Eustachian valve and auriculo-ventricular opening. A few veins open directly by small apertures into the right auricle.

The heart is supplied with *nerves* from the *cardiac plexus*. This is formed by branches from the sympathetic, the par vagum, and recurrent nerves. It gives off the coronary plexuses to accompany the arteries, and terminates in the muscular tissues of the heart.

SECTION II.
THE ARTERIES.

PULMONARY ARTERY, *Or.* from the circulus arteriosus of the right ventricle, which at this point dilates into a kind of *infundibulum*. At first in front of the aorta, between the two auricles, it crosses to its left side, and divides into its two pulmonary branches; the right, an inch and a half long, passes under the arch of the aorta, behind the descending cava, and in front of the bronchus; and the left, an inch long, in front of its descending portion and left bronchus, and behind the pulmonary veins. It is attached to the heart, 1st, by its inner tunic; and 2dly, by prolongations from the arterial zones, which are attached to the convex edges of its three festoons, and to the edges of their angular intervals. The fibrous pericardium covers the greater part of its two branches.

AORTA, *Or.* in the same manner as the pulmonary artery, immediately behind that vessel, in the anterior triangular interval formed by the ventricular zones. The arch commences from this point, which is opposite to the upper edge of the fourth costal cartilage of the left side, where it joins the sternum, and coming forwards ascends to the upper edge of the second costal cartilage on the right side; 2, it passes horizontally backwards on a level with the junction of the first with the second bone of the sternum to the left side of the body of the second dorsal vertebra; 8, it lastly descends, still inclining backwards, to the same side of the third or fourth dorsal vertebra, under the name of *thoracic aorta*, continues this course, coming more and more in front of the spine, enters the abdomen through the aortic opening of the diaphragm, and, finally, as the *abdominal aorta*, it passes to the body of the fourth lumbar vertebra, where it divides into the two *common iliac* and middle sacral arteries.

Relations of the arch. Within the pericardium, which reaches its transverse portion, the artery is, first, immediately behind the infundibulum of the right ventricle, and the pulmonary artery; it then gets to the right side of this vessel, and to the left of the vena cava descendens. Its transverse and descending portions form an arch over the root of the left lung; its transverse portion is partially covered by the left vena innominata; it rests on the bilfurcation of the trachea posteriorly. The phrenic and vagus nerves descend on each side, the left vagus a little in front of it. Its recurrent branch hooking round it where it is joined by the ductus arteriosus.

The primary Branches of the Aorta.

From the arch; 1, 2, coronary arteries, 3, *A. innominata*, 4, *left common carotid*, 5, *left subclavian*.

From the Thoracic portion; 6, *intercostales*, 7, *Bronchiales*, 8, *œsophageal*, 9, *mediastinal*.

From its abdominal portion; 10, *Phrenicæ*, 11, *cœliac*, 12, *sup. mesenteric*, 13, *inf. mesenteric*, 14, *capsula-ares*, 15, *renales*, 16, *spermatic*, 17, *lumbar*, 18, *common iliac*, 19, *sacra media*.

Art. *Coronaria* have just been described.

1st *set* from the convexity of the arch.

Art. Innominata,

from an inch to an inch and a half in length, *arises* from the summit of the arch, passes upwards, forwards, and to the right side, and divides opposite the right sterno-clavicular artic. into the right *com. carot.* and the *subclavian* trunks. Red, *behind*, the trachea and right pleura, *before* right vena innom., sternohyoid and thyroid m. upper bone of the sternum.

Right Common Carotid.[1]

This art. passes from its origin upwards, forwards, and a little outwards in the neck, along the inner edge of the sterno-mastoid nerve, until it arrives at the upper border of the thyroid cartilage, and divides into its terminal branches, the ext. and int. carotid arteries. In its first stage it lies in the ant. inf. triangle of the neck; it is here covered by the skin, platysma, and partially by the sterno-mastoid externally, the sterno-hyoid and thyroid internally; it is enclosed in the carotid sheath, lying internally to the vagus nerve and int. jug. vein, behind it are the sympathetic and recurrent nerves. In its second stage this art. lies in the ant. sup. triangle of the neck, and is covered only by the skin, platysma, and fascia; its other relations are similar to those in the first stage.

Left Common Carotid.

Arises from the arch of the aorta within the thorax, it is here covered by the left vena innom. and the origins of the sterno-hyoid and thyroid muscles, to its inner side is the trachea, to its outer the left subclavian and cone of the pleura, behind it is the thoracic duct and œsophagus. In the neck it is similar in course and relations to the right com. carotid, with the exception of its connection with the œsophagus and thoracic duct.

The right and left com. carotids differ in their origin, length, and a few of their relations, as just described. The terminating branches of the com. carotid art. are the external and internal carotid arteries.

EXT. CAROTID. Ascends inwards and forwards to the submaxillary gland; then backwards, imbedded in the parotid gland, parallel to and behind the ramus of the jaw, as high as the neck of the condyle. *Rel. ext.* integuments and platysma; the digastric and stylo-hyoid m. and ninth nerve; the seventh nerve, *int.* the stylo-glossus and pharyngeus m. which separate it from the *int. carotid Ram.* 10 in no.

Anterior Set.

Thyroid superior, Or. opposite the cornu of the os hyoides. It runs below this cornu, and the lingual nerve, and upon the sup. laryngeal nerve. *Ram.* 1, *hyiodeus*, cellular membrane between the stylo-hyoid muscle; 2, *superficiales*, sheath of the carot., glands, and sterno-mastoid muscle; 3, *laryngeus*, larynx, epiglottis; a branch descends on the cricothyr. memb.; 4, thyroideus, generally three, to the thyroid gland.

Lingualis, runs, 1st, horizontally above the cornu of the os hyoides; 2d, vertically, to the base of the tongue; and 3d, horizontally, as the *Ranine artery*, to its tip. *Ram.* 4. 1, *hyoideus* supra hyoid muscles, epiglottidean glands; 2, *dorsalis linguæ*, dorsum of the tongue, tonsil, pal. arches, &c.; 3, *sublingualis*, sublingual gland, mucous membrane; 4, *Ranine*, along the outer side of the genio gloss. m. to the tip of the tongue.

Facialis, upwards and inwards to the submax. gland, forwards on the upper edge of the gland, and upwards from the ant, edge of the masseter to the labial commissure, ala nasi and inner canthus of the orbit. *Ram.* 6. 1, *Inf palatine*, ascends between the stylo gloss. and pharyngeus to the sup. constrictor tonsil, velum, &c. 2, *Tonsillaris*, between the stylo gloss. and int. pteryg. m. to the tonsil; 3. *Glandulares,* to submaxillary gland; 4, *Submentalis* runs close under the jaw, to the muscles at the symphysis menti; 5, *abialis. inferior,* lower lip and chin; 6, *Coron. inferior,* mucous membrane of the lower lip and labial glands; 7, *Masseteric,* masseter, buccinator; 8, *coronaria* sup., mucous membrane of the upper lip and labial glands; 9, *nasi lateralis,* side of the nose in numerous anastomoses; 10, *angularis* lacrymal sac, orbicularis muscle.

Posterior Set.

Sterno-mastoid, to that, and to the adjoining muscles and glands.

Occipitalis, ascends parallel to, and concealed by, the post. belly of the digastric; passes between the transverse process of the atlas and mastoid process; and then upon the recti beneath the other cervical muscles which it lastly perforates at the centre of the sup. transverse ridge. *Ram.* 1, *mastoid;* 2, *post meningeal;* 3, *cervicalis descendens,* which descends between the splenius and complexus; 4, *parietal,* parietal foramen; and 5, *terminal,* upon the back of the cranium.

Posterior auris, Ram. auricular, muscular, parotides, and *stylo-mastoid,* which passes through the Fallopian canal, to the tympanum and labyrinth.

Inner Set of Branches.

Pharyngea Ascendens ascends deeply by the side of the pharynx; *Ram. pharyngeal,* to the pharynx, tonsils, and palatine arches, *post. meningeal,* through the post, foramen lacerum; *ant. meningeal,* ant. foramen lacerum; *Eustachian,* to that tube, and to the tympanum.

Terminal Set of Branches.

Trans. faciei, runs a little below the zygoma to anastomose on the face with the other facial branches.

Internal maxillary passes 1, inwards behind the

neck of the jaw; 2, upwards and inwards, in a space bounded by the two pterygoidei m., the buccinator and the ramus of the jaw; 3, horizontally, forwards between the heads of the pteryg. ext. across the pteryg. maxillary fossa to the infra orbitar canal *Ram.* first part of its course; 1, *Art. meningea media magna,* through the foramen spinale; ramus hiatus Fallopii to the facial nerve; 2, *Infra dental, Ram.* mylo-hyoid, to that muscle, and labial glands.

Second Order.

3, *Deep temporal,* two to the deep surface of them, 4, *Massetericæ;* 5, *Pterygoidei;* 6, *Buccales;* 7, *Supra dental,* roots of the post. teeth, and antrum.

Third Order:

8, *Infra orbitar* (terminal) antrum, ant. alveoli, and face; 9, *palatina descendens,* pos. palatine canal, palate, *ram. incisivus;* 10, *Nasalis,* spheno-palatine foramen, and nose, *ram. nerve. Cotunnius;* 11, *videan,* pterygoid foramen, to accompany the nerve.

Temporalis; Ram. 1, *auricular,* capsule of the jaw; 2, Terminal, *temporales, anterior, media et posterior.*

INT. CAROTID, runs upwards, vertically, to the foramen caroticum, forwards along the canal, tortuous, through the cavernous sinus, vertically, by the side of the anterior clinoid processes. *Rel. Int.* side of the pharynx, *tonsil,* symph. nerve; *ext.* pneumogastric glossopharyngeal, and ninth nerves; int. jugular vein; *post.* vert. column; *ant.* stylo gloss. et. pharyngeus m. Internal to all the nerves in the cavernous sinus. Ram.

A, *Opthalmic* passes with the nerve through the foramen opticum. In its passage through the orbit it lies 1, on the outside of the nerve, 2, on its upper part, 3, on its inside. *Ram. first part of its course:* 1, *Lacrymal* to the gland, palpebra, some pierce the os mala; 2, *Centralis Retina,* centre of the optic nerve and vitreous humour, to the posterior surface of the lenticular capsule.

Second Part of its Course.

3, *Supra orbatilis*, through the supra orbitar notch to the forehead; 4, *ciliary posterior*, 15 to 20, through the sclerotic coat, to the choroid coat, ciliary processes, and to the larger circle of the iris; 5, *ciliary long.* two in no., *anas.* in circles on the post. surface, and smaller circle of the iris; 6, *muscular*, superior and inf. *Ram.* ciliary anterior, to the larger circle of the iris, and conjunctiva.

Third Part of the Opthal. Art.

7, *Ethmoidal posterior;* through the post. For. Eth. to the dura mater, and nasal fossæ; 8, *Ethmoidal anterior;* ant. For. Eth. to the frontal sinus, nares; 9, *Palpebral, superior et inferior.*

Termination of the Opth. Art.

10, *Nasal*, anastom. with the term. branch of the facial; 11, *Frontal*, to the inner part of the forehead.

Art. communicans post. (Willis) runs backwards, along the outer side of the pituitary gland and corp. mamil., to join the post. cerebral artery, which is a branch of the Basilar.

Choroid, backds. and outds. along the optic fascic., enters the lat. vent. through the great fissure, and is lost in the choroid plexus.

Ant. cerebral, anastomoses with its fellow, by means of the *art. communicans ant.* and then runs, from before backwards, on the upper surface of the corpus callosum; supplies the hemispheres.

Middle cerebral, in the fissure of Sylvius, to the ant. and middle lobes.

Cereb. Branches from the Art. Basil.

Post Cerebral, to the Thalami, Tuberc. Quadrigemina. etc. It is joined by the *communicans* of Willis.

Ant. Cerebellar super. surface of the cerebellum.
Ram. Int. auditory, through the int. meatus to the internal ear.

Post. Cerebellar, sometimes arises from the vertebral; passes between the origins of the ninth pair of nerves, and just in front of those of the eighth, to the post. part of the cerebellum.

Subclavian Artery

extends from its origin to the lower edge of the first rib; it runs, 1st, upwards and outwards to the scaleni; 2d, horizontally between them; 3d, downwards and outwards to its termination, *Rel. before it reaches the scaleni.*

The *Right*, 1, *Ant.* sterno. clavic. artic.; *parallel* to the subclavian vein where it joins the jugular; 2, *crossed* by the phrenic and vagus nerves; 3, *post.* recurrent nerve; pleura: surrounded by filaments of the sympathetic nerve.

The *Left*, differs, 1, *Ant.* its *crossed* by the subc. vein; 2, the phrenic and vagus are *parallel* to it; 3, it is nearer the clavicle than the spinal column; 4, in more extensive contact with the cone of the pleura and lung.

The thoracic duct ascends between it and the left carotid, it is larger than the right, and is more deeply seated in its first course.

In the root of the neck, before passing behind the scalenus antic. muscle, the subclav. art. are covered by the skin, platysma, sterno-mastoid and sterno-hyoid muscles and cervical fascia; this is the first stage.

Second Part.—Between the scaleni; the brachial plexus above and posterior, and the first rib below, the cone of the pleura behind; the ant. scalenus separates it from the vein. *Third Part.*—Lies in the posterior inf. triangle of the neck covered by the skin, platysma, and a quantity of loose cellular tissue, the ext. jug. vein descends in front of it, and its own vein lies ant. and inf. to it; the brachial plexus lies above and behind, and the first rib immediately posterior to it. *Ram.*

First Part of its Course.

Art. vertebral, (always the first branch,) runs through the transverse foramina of the cervical vertebræ, from the sixth (sometimes fourth) to the second, passes outwards to reach that of the first, and winds horizontally backwards, in a groove behind its artic. process; pierces the post. occip. atloidean lig. beneath the suboccipital nerve, and, proceeding forwards and upwards through the foramen magnum, inclines to its fellow on the ant. surface of the medul. oblongata. It lastly, unites with it at the pons to form the *Art. Basilaris. Ram.* 1 and 2, anterior and posterior *Medulla spinalis.* 3, *Basilar art.*, extends the whole length of the pons Varolii.

Internal mammary arises opposite the last, descends in the thorax behind the costal cartilages, close to the sternum, to the xiphoid cartilage. *Rel.*, crossed by the phrenic nerve which then lies on its inner side. *Ram.* 1, *ant. intercostal,* 5, to the five superior intercostal spaces; 2, mediastinal, thymicæ, glandulares, muscular, &c.; 3, *Comes, nervi phrenicæ,* runs with the nerve to the diaphragm. It terminates in 4, *musculo-phrenicæ,* diaphragm and inf. intercostal spaces; 5, *abdominales,* anas. with the epigastric and lumbar art.

Thyroid. an axis of three branches; 1, *Thyroid inf. or ascendens,* ascends behind the carotid sheath to the thyroid gland; 2, *cervicalis ascendens,* ascends on the ant. scalenus, parallel to the phrenic nerve; 3, *supra scapularis,* or *transversalis humeri,* to the supra scapular notch. *Ram.* supra acromial, supra spinalis, and infra spinalis.

Transversalis colli, across the neck above the trans. humeri. *Ram.* 1, *cervical superf.* to the fascia, lymphatic glands, &c. 2, *scapularis post.* (terminal) post. angle, crista and inf. angle of the scapula.

Second Part of the Subclav. Art.

Cervicalis profunda, passes between the 6 and 7 cervical transverse processes, and ascends deeply in the spinal groove; *anas.* with the occipital and vertebral art.

Intercostalis sup. descends on the outer side of the first thoracic ganglion; *Ram.* 2, one to the first, and one to the second intercostal space.

Axillary Artery

Extends from the infer. edge of the first rib to the lower margin of the latissimus dorsi and teres muscles; it is divided into three stages by the pectoralis minor, which crosses in front of it. *Rel. Ant.* Pect. major above and below; pectoralis minor in middle *post.* a cellular space, then teres major and latissimus dorsi m.; *Ext.* coracoid process, subscapularis m. and head of the humerus; *int.* first intercostal space, skin of the axilla. The vein, first at some distance on its inner side, gets nearer and more anterior to it in descending. The brachial plexus. first on its outer side, gradually embraces it; the inner root of the median, the *int. cutan.* and ulna nerves, are given off on its inner side; the *ext. cutan.* and outer root of the median on its inner side; the musculo spiral, and circumflex nerves, behind it. *Ram. At the upper edge of the pectoralis minor.*

Thoracica acromialis (constant.) *Ram.* Pectorales acromialis, deltoidean; a branch separates the deltoid from the pectoralis major m.

Thoracica suprema (irregular,) ramifies between the pectoral muscles; *anas.* with the int. mammary art.

Thoracica alaris (irreg.) to the cel. tissue and glands in the axilla.

Below the pect. minor.

Thoracica longa (const.) descends parallel, and close to the lower edge of the pector. minor m., between the p. major and serratus magnus, to the side of the thorax.

Subscapularis runs along the lower edge of the subscapular m. accompanied by the subscap. nerve, to the inf. angula of the scapula. *Ram.* 1, *Dorsalis scapulæ*, through the triangular opening formed by the long head of the triceps, teres major, and edge of the scapula, to the subscapular fossa.

Post. circumflex ⎱ encircle the surgical neck of the
Ant. circumflex ⎰ humerus. *Ram.* 1, *Deltoidean;* 2, *artic.* which ascends in the bicipital groove.

BRACHIAL ARTERY.

Extends from the termination of the preceding to the bend of the arm, in a line from the centre of the axilla to the middle of the latter; *Rel.* 1, *in the arm:* *Ant.* inner edge of the coraco brachialis and biceps, which overlap it; *Post.* triceps, brachialis anticus and musculo spiral nerve. *Int.* ulner nerve and inf. profund. art. *Ext.* coraco brachialis, cellular interval between the brachialis antic. and biceps. The median nerve, first on its outer side, crosses at about its middle to lie on its inner side; 2, *Rel. at the elbow; Ant.* expansion of the biceps, skin, basilic vein, *Post.* middle of the brachialis anticus, *Int.* median nerve, *Ext.* tendon of the biceps, *Ram.*

Profunda Sup. descends in the spiral groove round the posterior surface of the humerus to *anas.* with the radical recurrent in the groove between the brachialis antic. and supinator longus, *Ram.* 1, descends to the inner part of the elbow-joint; 2, to accompany the deep branch of the spiral nerve; 3, musculares.

Profunda Ins. descends on the inner side of the arm along with the ulnar nerve to *anas.* with the post. ulnar recurrent, between the int. condyle and olecranon process. *Ram.* musculares.

Anastomotica Magna, anas. super. with the inf. profunda, and *inf.*, with the ant. and post. ulnar recurrent.

Art. Nutritia humeri; i. e. Terminal, Radial, and Ulnar arteries.

The brachial art. is superficial throughout, being covered only by the skin and fascia, and in the centre by the inner edge of the biceps, nervous filaments from the int. cut. and nerves of Wrisberg lie over it; it is accompanied by the brachial vein which lies to its inner side. Below the bend of the elbow it sinks into a triangular fossa, bounded by the brach. antic. above, the pronator teres internally, and the supinator longus externally, and divides opposite the coronoid process of the ulna into its terminal branches, the Radial and Ulnar arteries.

Radial Artery

Extends from the elbow to the wrist, in a line from the middle of the elbow-joint to the root of the thumb. *Rel. Ext.* Supinat. Rad. Long.; and for some distance the radial nerve. *Int.* Pronator teres, and flexor carpi radialis. *Post.* attachments of the supin. brevis and pron. teres; flexor long. poll.; and ant. surface of the radius. *Ant.* above supin. long., below fascia and skin. *Ram.* 1, *Recurrent*, anas. with the sup. Profunda; 2, *Superficialis Volæ*, runs over the annular lig. through the short muscles of the thumb to join and complete the superficial palmar arch; 3, *Ant. Carpi Radialis.*

The vessel now passes backwards beneath the two first extensors of the thumb to the first interosseous space, which it traverses above, or through the fibres of the abductor indicis m. Here it may be felt in a space bounded by the two first extensor tendons of the thumb, and that of the third, *Ram.* at this stage. 4, *Dorsalis carpi*, *Rad.* carpus and wrist-joint; 5, and 6, *Dorsalis Pollicis*, dorsal edges of the thumb; 7, *Princeps* vel *magna pollicis*, runs between the abd. indicis and abd. pollicis, divides at the first phalanx of the thumb into its two margin branches; 8, *Radialis Indicis, anas.* with the superf. palmar arch, and become a digital artery. The terminal branch, or *Palmaris profunda*, crosses the carpal heads of the metacarpal bones, covered by all the soft parts, and on that of the little finger joins the Ram. communic. of the ulnar art. and thus forms the deep palmar arch. The *Ram.* of this arch are five in number. They supply, the interossei, and at the cleft of the fingers, join the branches of the superficial arch before they bifurcate.

Ulnar Artery

Descends, with a curve towards the inner side of the arm, from the elbow to the inner side of the pisiform bone. *Rel.* at the upper third of the arm it lies between the two layers of flexor muscles; it afterwards rests, with the ulnar nerve on its outside, on the flexor profundus. At the middle third of the arm it is covered by the adjoining edges of the flexor sublim., and flex. carpi ulnaris; at the lower third by the skin, and fascia: *Ram.*

1—2, *art recurrent*, ant. et post. *anas.*, the first with the anastomotica magna and the inf. profunda; the second with the former vessel and the sup. profunda; 3, *interossea*, runs downwards and backwards to the interosseous space, where it gives off two *recurrent* branches, and the *ram. interos. post.*; it then descends on the ant. surface of the ligament, and at the upper edge of the pronator quadratus divides into *ant.* and *post. carpal* branches, the former anas. with the deep palmar arch, and the latter with the ram. dorsalis carpi. The *ram. interos. post.* perf. the upper part of the lig., gives off a post. interos. recurrent, and

14

descends as far as the dorsum of the carpus. The *recurrent* branches *anas.* with branches of the radial art.; 4, *descend. muscul.* vel *nervi mediani* to the superficial flexors; 5, 6, *r. carpi ulnaris*, ant. and post.

The ulnar artery now passes from the inner side of the pisiform bone, over the annular lig., gives off the *communicans profund.* to complete the deep palmar arch, and then as the *r. palmaris superf.* it arches across the hand, beneath the palmar fascia and skin, and, between the ball of the thumb and index finger, joins the superf. volæ and rad. indicis, to form the *superf. palmar arch.* Four digital branches are given off, each of which first anas. with a branch from the deep arch, and then divides into two collateral branches; they supply the edges of the three last fingers and the ulnar side of the index.

The digital vessels and nerves run along the lateral surfaces of each finger, and nearer its anterior surface; they terminate in a free anastomosis in the expanded extremities of the fingers, and endow these with common sensibility, and the special sense of touch.

THE THORACIC AORTA

is situated in the post. mediastinum, and extends from the left side of the fourth dorsal vertebra to the aortic opening of the diaphragm. *Rel. ant.* the left pleura, root of the left lung, œsophagus, the left auricle of the heart, and pericardium, *post.* left side, and afterwards the fore part of the spinal column: owing to this inclination, the œsophagus and vagi, which above are placed on its right side, cross it to lie, below, rather to its left side; behind it, also, are the left intercostal veins, going to join the azygos. The splanchnic nerves lie on each side of it, and the thoracic duct and azygos vein on its right side. *Ram.*

Bronchials, generally four in number, two on each side; their branches twine around the bronchi to supply the substance of the lung; *anas.* with the pulmonary artery (Ruysch, Cowper, Waller,) with the pulmonary veins (Harrison.)

Œsophageal, from one to four or five to the œsophagus; they *anas.* with other art. by ascending and descending branches.

Intercostales, from eight to ten on each side, they run in the eight or ten inf. intercostal spaces, and *anas.* with the *int. mammary a.* Each gives off a *ram. postic.*, which passes between the spine and costo-transverse lig. to supply the *cord*, and deep dorsal muscles.

THE ABDOMINAL AORTA

descends, inclining slightly to the left side, from the diaphragmatic opening to the body of the fourth or fifth lumbar vertebra. *Rel. ant.* lesser omentum, stomach, solar plexus, vena porta, mesenteric vessels, pancreas and duodenum, transverse mesocolon, root of the mesentery, and small intestines. *Right side*, vena cava, thoracic duct; both these vessels are behind it at their origins. On each side lies the sympathetic nerve. Post. the pillar of the diaphragm and lumbar vertebræ. *Ram.*

Diaphragmatic, right and left, they encircle the diaphragm, and join at its ant. edge; anas., with lower intercost. arts.

CŒLIAC AXIS, a short trunk projecting horizontally from the aorta immediately below its diaphragmatic opening. *Rel.* the pancreas and vena porta below it; the lob. Spigelii on its right; the renal capsules and semilunar ganglions on each side; enveloped by the solar plexus and covered by the stomach. It is an axis of three branches.

a, *Gastric* or *coronar. ventriculi*, runs to the cardiac orifice of the stomach, and divides into, 1, *ram. super.* to the œsophagus, and larger end of the stomach; 2, *ram. infer.* follows the lesser curvature, *anas.* with the sup. pyloric art. of the hepatic.

HEPATIC, runs enclosed within the smaller omentum, 1st, transversely towards the pylorus, and 2dly, upwards and forwards to the transverse fissure of the liver. *Ram.* 1st stage, 1, *sup. pyloric*, to the pylorus, *anas.* infer. branch of the gastric; 2, *gastro-duodenalis*, accompanies the ductus communis choledochus, and between the duodenum and pancreas, divides into *pancreatico duodenalis*, and *gastro-epiploica dextra.* The former supplies the corresponding organs, the latter runs along the larger curvature of the stomach, and *anas.* with the gastro-epiploica sinist., a branch of the splenic.

Terminal Branches.

3, *Hepatica sinistra*, left lobe of the liver; 4, *hepatica dextra*, to the right lobe; from this is derived the *ram. cystica* to the gall bladder.

SPLENIC, five or six inches long; it runs with the vein to the left side, along the upper and post. part of the pancreas, and terminates by five or six br. in the spleen. Ram. 1, *pancreaticæ parvæ*, ramusculi to the gland; 2, *pancreatica magna*, runs from left to right along with the duct; 3, *vasa brevia;* five or six large, but short, branches arising from the trunk, or its splenic divisions, to ramify in numerous *anas.* on the larger end of the stomach; 4, *splenicæ*, five or six in number, to the spleen; 5, *gastro-epiploica sinistra*, runs from left to right on the larger curvature of the stomach to *anas.* with the right corresponding vessels from the hepatic.

SUPERIOR MESENTERIC, arises from the aorta at the lower edge of the pancreas, which separates it from the cœliac, and above the transverse portion of the duodenum. It descends, forming a slight curve, to the *right* iliac fossa. Ram. 1st, *from its convexity.*

a, *Mesentericæ*, fifteen or twenty branches, they *anas.* with each other in a series of arches, which become smaller and more numerous as they approach the small intestines, to which they are finally distributed.

2d, *from its Concavity.*

b, *Colica media*, to the transverse colon, *anas.* on the left with the *colica sinist.*, on the right with the *colica dextra.*

c, *Colica dextra*, to the ascending colon, *anas.* above with the *colica media*, below with the *ilio colic.*

d, *Ilio colic*, to the cæcum, valve, &c., *anas.* above with the colica dextra, below with branches from the inferior mesenteric art.

INFERIOR MESENTERIC, arises about an inch above the bifurcation of the aorta, and descends towards the left iliac fossa. *Ram.*

a, *Colica sinistra*, to the descending colon, *anas.* above with the colica media, below with the art. sigmoidea.

b, *Art. sigmoidea*, to the sigmoid flexure, *anas.* above with *colica sinistra, ram.* to the psoas, iliacus m. and ureter.

c, *Hæmorrhoidalis superior*, descends in the mesorectum to the rectum, and at about its middle divides into two branches, which descend on the sides of the organ, some of their ramusculi even reaching the anus; *anas.* with the middle and inf. hæmorr. art.

All the foregoing branches of the abd. aorta are accompanied each by a plexus of nerves, named after each vessel, and derived from the semilunar ganglion, solar plexus, and sympathetic nerve. Their blood is returned by the vena porta, which passes through the liver before pouring its blood into the inf. vena cava.

Capsulares, large in the fœtus, supply the suprarenal capsules.

RENALES, arise between the two mesenteric arteries, the right, which is the longer, passing behind the vena cava. Their branches, four or five in number, run between the mammillæ of the kidney, *anas.*

in arches, and then terminate in its cortical substances. Some branches reach the capsule.

Spermaticæ arise immediately below the former, the right frequently from the right renal. They descend, crossing the psoæ m. and ureter on each side, and at the inner ring join the vas deferentia, which guide them to the testicles. In the female they run to the ovaria.

Ram. in the male, a, ram. to the psoæ, ureter, and adjoining tissue; b, six or seven to the epididymis and body of the testis.

In the female, a, ram. to the ovaries, especially to the uterus, where they *anas.* with the uterine arteries; b, ram. accompany the round lig.

Lumbales, five in number, corresponding to the vertebral spaces. They divide into, a, *ram. spinales* to the cord, and through a nutr. foramen to the body of each lumbar vertebra; b, *post. muscular;* c, *circular* or *abdominal,* analogous in their course to the intercostales with which they anastomose.

Terminal Branches of the Aorta.

Sacra media (caudal artery of animals) descends on the sacrum and coccyx; it supplies the rectum, and *anas.* at each division of the above bones with the lateral, sacral, and hæmorrhoidal arteries.

The abd. aorta at its bifurcation lies in front of, and a little to the left side of, the fourth lumbar vertebra; it is here nearly opposite the umbilicus, and has the vena cava to its right side, the sympathetic on either side; it bifurcates into the

ILIACÆ COMMUNES

extend from the fourth lumbar vertebra to the *sacro iliac artic.* *Rel.* crossed by the ureters and spermatic vessels, resting above upon the body of the last lumbar vertebra, below against the inner edge of the psoas m. The two corresponding veins lie beneath and on their inner sides, and the left in joining the right to form the *vena cava ascend.*, passes beneath the right iliac artery.

Ram. to the psoœ, ureters, &c., they sometimes give off the sacra media, and last lumbar arteries, and even the renales, and spermaticæ (Cruveilhier;) b, ram. terminal, *iliacæ extern et intern.*

Internal Iliac,

About an inch and a half long, descends vertically (bending slightly forwards) from the sacro iliac joint to the upper edge of the sacro sciatic notch. Many of its *branches* sometimes arise from a common trunk.

Anterior Set of Branches.

a. *Umbilical*, large in fœtal life, they soon dwindle after birth into fibrous cords, which may be traced along the sides of the bladder to the umbilicus; near their origins, where they still remain pervious, some of the principal vesical branches arise.

b. *Vesicales*, four or five, derived principally from the umbilical, and from the middle hæmorr. obturator, vaginal, and uterine arteries. *Ram.*, 1, *infer.* vesic. semin., *prostate gland*, and neck of the bladder; 2, *media*, along the uterus to the sides of the bladder; 3, *super.*, ant. surface of the bladder.

c. *Obturator*, may arise from the femoral or epigastric art., when it winds round the infer., or the super. and inner margins, of the femoral ring. It runs parallel to, but below, the ext. iliac vessels, accompanied by its nerve (which is above it,) to the obturator foramen. *Ram.* 1, a cross branch, by which it joins its fellow behind the pubis; 2, *vesicales musculares, iliacæ* to *anas.* with the circumflexa ilia. Having traversed the obturator foramen, its branches are, 1, *Ram. intern.* runs round the inner half of the obturator foramen, between the two obt. m., which it supplies, and *anas.* with the *ram. extern.*; a *branch* descends between the two first adductor m. 2, *ram extern.* between the obt. m. round the ext. half of the foramen, *anas.* between the neck of the femur and quadratus muscle with the ischiatic art.; a *branch* enters the joint through the notch; *other br.* supply the muscles.

Int. Branches.

d. *Middle hæmorrhoidal*, to the ant. part of the rectum, *anas.*, above and below, with the other hæmorrh. branches.

e. *Uterine*, runs in the broad lig. to the neck of the uterus, on which it ascends to *anas.* with the spermatic art.

f. *Vaginales*, surround the vagina.

Posterior Branches.

g. *Ileo-lumbar*, divides behind the psoas. m. into, 1, *ram. ascend.*, *anas.* with the last lumbar art.; 2, *transversales*, superficial and deep, to the iliacus m.; 3, the *nutritious art.*

h. *Sacra lateralis*, descends in front of the sacral foramina. *Ram.* to the spinal cord, and to the muscles on the back of the sacrum; *anas.* with the sacra media.

i. GLUTEAL, escapes between the upper edge of the sciatic notch and pyriformis m., and immediately divides into, 1, *ram. superf.*, runs between the g. max. et medius m.; 2, *r. profund.*, between the g. medius and minim.; it sends a branch along the *g. min.* to the spin. p. of the ilium, a second across this m. to the *g. medius*, a third through the *g. min.* to the artic. capsule.

Inf. Branches.

k. ISCHIATIC, escapes from the pelvis, between the lower edge of the pyrif. and sciatic lig. *Ram.* to the gluteus max., lev. ani, and coccygeus m.; 2, *comes nervi ischiatici.*

1. INTERN. PUDIC, descends upon the pyriformis m., internal to the former vessel, with which it escapes from the pelvis; 2, it winds over the spine of the ischium, beneath the larger sciat. lig., and enters the *smaller* sciatic notch; 3, it ascends on the inner side of the tuber ischii and rami ischii et rubis, but not within the pelvis, to the arch of the pubes, covered and protected by a prolongation of the large sacr. sciatic lig., as far as the crus penis.

Ram.—First Stage.

1. *Vesical, prostatic,* and *sometimes* the middle hæmorrhoidal.

Second Stage.

2. *Anastomotic, anas.* with the ischiatic and int. circumflex art.; 3, *musculares,* to the flexors at the tuber ischii.

Third Stage.

4. *Ext. hæmorrhoidales,* two or three to the anus; 5, *perineal* (superf.) pierces the middle fascia of the perinæum to get into the space between the accel. urinæ and erect. penis m.; it ascends to the septum scroti, which, and the superficial parts of the perineum, it supplies; 6, *transvers. perinea* (superf.) to the perinæum and anus; 7, *corp. bulbosi* (deep transverse perinæal,) runs inwards behind the middle perinæal fascia, to the bulb, Cowper's glands, &c.

Terminal Branches.

8. *Corp. cavern.* to those bodies; 9, *dorsalis penis* the two latter *anas.* by a cross branch, or unite into a single vessel; then pierce the suspensory ligament, and run on the dorsum of the penis as far as the *glans.*

The internal iliac art. is accompanied by its vein, which lies anterior and internal to it; behind it is the sacral plexus, with whose branches those of the art. interlace, and still more posteriorly the origins of the pyriformis musc.

External Iliac

extends from the sacro iliac joint, to the inf. edge of Poupart's lig., in a line leading from a point half an inch external to its middle, to the umbilicus. *Rel.* bound down by a process of the iliac fascia; it lies, first, to the inner side, and then on the fore part of, the psoas muscle, to which it is connected by a layer of the iliac fascia; the vein lies on its inner side; close to Poupart's ligament it is crossed by the ilio scrotal nerve. It is surrounded by numerous lymphatics, and covered by their glands. *Ram.*

a. *Epigastric, Or.* from any point of the inferior inch and a half of the artery. It first reaches the inner margin of the *int.* inguinal ring, and then ascends inwards between the peritoneum and fascia transversalis; it soon enters the sheath of the rectus m., in which it term. about an inch on the outer side of the umbilicus. *Rel.*, the spermatic cord lies in front, the vas deferens looking round it to descend to the vesic. semin. *Ram.*, 1, *cremasteric* to that structure and the scrotum; to the round lig. in the female; 2, *anastomotic, anas.* with its fellow and the obturator art.; 3, terminal *anas.* in the rectus, with the *int. mammary art.*

b. *Circumflexa ilii* (internal,) outwards, along Poupart's lig. and crista ilii; *anas.* with the last lumbar, and the ilio lumbar art.

Femoral Artery

extends from the termination of the ext. iliac to the lower third of the thigh, in a line leading from the middle of Poupart's lig. to the inner side of the internal condyle of the femur. *Rel.* at the *upper third* of the thigh it lies in a triangle (Scarpa's) formed by

Poupart's lig., the sartorius, and add. longus, covered only by skin, fascia, and glands; at the *middle third* it is covered by the sartorius. It rests on the tendon of the psoas and iliacus m., pectineus, and adductor longus m. The vein and artery are enclosed in a sheath from the fascia lat. This sheath becomes a strong fib. canal, where the vessels rest on the tendon of the adductor longus, beneath the sartorius muscle. The vein, at first on its inner side, gets more and more behind it as it descends; the *ant.* crural nerve sends its saphenous branch into the upper part of its sheath: the vena saphena runs between the skin and fascia, a little to its inner side. *Ram.*

a, *Superficial epigastric* and *superficial circumflexa ilii*, to the abdominal integuments; b. *superf. pudic*, two or three running beneath the fascia to the scrotum; c, *muscular superf.*, to the iliacus, psoas, and rectus m.

d. PROFUNDA, *Or.* generally an inch or an inch and a half below Poupart's ligament. It is placed behind, and separated from the femoral by the *profun.* and *fem. veins*, and *ad. long. Ram.* 1, *ext. circumflex*, passes between the sartorius and rectus m., divides into a *transv. branch* to the hip-joint and muscles, and a *ram. descend.*, which runs between the rectus and crureus, as far as the knee; 2, *intern. circumflex.*, passes backwards, between the pectineus and psoas m.; *ram. superior* to the digital fossa; *ram. transv.* between the quad. fem. and add. long. to flexor m. 3, *perforantes*, three in number—1, *super.*, backwards between the add. brevis and pectineus; 2, *media.* through the add. brevis and magnus; 3, *infer.*, through the add. magnus, at the upper edge of the add. longus; 4, *r. terminal*, passes behind the add. longus, and perforates the add. mag. These vessels supply the flexor muscles on the back of the thigh, and form a chain of anastomoses by their ascending and descending branches; 5, *anastomotica magna*.

Sometimes a large artery; descends towards the inner condyle, and there enters into numerous anastomoses.

Popliteal Artery

extends from the femoral opening in the add. magnus to the inf. edge of the popliteus m., in a line descending along the centre of the back part of the knee-joint. It rests in a mass of fatty tissue in a diamond-shaped space, formed above by the divergence of the ham-string m., and below by the convergence of the two heads of the gastrocnemius. The vein is posterior and external to the art., and the popliteal nerve is in similar relation with the vein. *Ram.*

a. *Muscular super.*, two or three to the ham-string mus.

b. *Artic. super. ext.*, winds around the femur, beneath the biceps tendon, and divides into *r. superf.*, which ramifies on the patella and *r. profund.* to the synovial membrane and femur.

c. *Artic. super. int.* winds round the inner side of the femur, beneath the inner ham-strings.

d. *Artic. inf. ext.*, runs along the edge of the ext. semilunar cart.

e. *Artic. inf. ext.*, winds around the inner side of the tibia. These three art., like the first, divide into superf. and deep branches, which anas. the former on the patella, the latter about the joint.

f. *Artic. media* (Azyga,) perforates the lig. postic. to supply the joint.

g. *Muscular infer.*, (Surales,) to the gastrocnemius. The popliteal *art.* now termin. into the *Ant.* and *Post. tibial art.*

Art. Tibialis antic., runs between the heads of the tibialis post. m., and above the upper edge of the interosseous ligament; descends on its ant. surface, and terminates in the fissure between the tarsal heads of the first and second metatarsal bones. *Rel.* between the exten. comm. and tibialis antic. m. above, between the former and exten. prop. pol. below. The nerve runs on its outer side. *Ram.*

a. *Recurrent,* anas. with the int. artic. a.

b. *Musculares,* adjacent muscles.

c. *Internal malleolar,* and d. *ext. malleolar,* ramify on those processes.

f. *Metatarsal,* which gives three dorsal interosseous branches.

g. *Pollicis,* supplies the two sides of the first toe, and inner side of the second.

h. *Communicans (terminal,)* joins the ext. plantar artery in the sole of the foot, to complete the plantar arch.

POST TIBIAL ART. descends between the superficial and deep flexor muscles to a point midway between the int. malleolus and os calcis; in its course, it rests successively on the tibialis posticus, flexor communis, and, lastly, on the bone, where it is placed between the sheath which encloses the tendons of the two last named muscles, and that of the flexor prop. pollicis. *Ram.*

a. *Peroneal,* descends between the inner edge of the fibula and attach. of the flexor pollicis m. to the ext. malleolus. *Ram.* 1, *ant. peroneal,* pierces the interosseous ligament to reach the tarsus; 2, *terminal,* to the outer malleolus, &c.

b. *Internal plantar art.* runs upon the upper surface of the adductor pollicis m. to the integuments of the great toe; sometimes it joins a branch from the next vessel, to form a superficial plantar arch.

c. *Ext. plantar art.* runs first to the tarsal end of the fifth metatarsal bone, and then across the foot next the interosseous muscles, to join the *communicans* of the ant. tibial, and form the principal palmar arch. *Ram.* 1, *perforantes,* three in number; they pierce the interossei to join the interosseous branches of the metatarsal artery; 2, *digital,* four in number, arise from the convexity of the arch, and supply both sides of the three last toes, and outer side of the second.

Collateral Circulation.

The most remarkable feature in the arrangement of the blood-vessels is their anastomoses. The arteries and their branches enter into combinations with each other at every possible opportunity.

A. Every artery anastomoses with its fellow, excepting some supplying the viscera, and those in the extremities.

B. Every artery anastomoses with the vessel immediately above and below it.

C. Every artery anastomoses with itself by means of its branches, which follow the general rules of anastomoses.

As interesting examples of the first may be mentioned:—The anas. between the ant. cerebral art. of the int. carotids by the ant. communicans.

2. Between the obturator arteries, before their exit from the pelvis, by a transverse branch.

3. Between the dorsal arteries of the penis, by a transverse branch.

4. Between the vertebral arteries, which unité to form a single vessel, the basilar.

5. Between the epigastric art., by a transverse branch.

6. Between the ext. carotids, by nearly every one of their branches, which also form the most remarkable examples of the second rule, &c., &c.

7. The intercostals and upper lumbar arteries encircle the trunk.

Examples of the second rule.

1. In the inf. extremity, all the branches of the ext. iliac and femoral arteries form a posterior or principal, and anterior or minor collateral circulation, by a chain of anastomoses along the limb, each

of which is sufficient to carry on the circulation when the principal trunk is obliterated.

2. In the superior extremity, a chain of anas. may be traced along the margins of the scapula, inner edge of the arm, &c.

3. From the angle of the eye, along the nose, mouth, face, and median line of the neck; and, by means of the int. mammary and epigastric arteries, along the chest and abdomen, even to the principal art. of the inf. extremities.

4. Along the œsophagus, stomach, and intestines, an anastomotic chain may be traced from the pharynx to the rectum.

5. Along the spinal cord, by means of the spinal arteries which hold numerous anastomoses through the intervertebral formina with the vertebral, intercostal, lumbar, and sacral arteries.

As interesting examples of the third rule may be mentioned,

1. The branches of the external and internal carotids; the former in the brain, the latter in the neck.

2. The branches of the cœliac axis, round the stomach; the coronary arteries, round the heart; the ciliary arteries round the iris, between its larger and smaller circles; the branches of the super, and inferior mesenteric arteries, especially the former: the branches of the articular arteries, especially those of the knee-joint. The digital branches in the hand and foot, the branches of the obturator artery round the margin of the obturator foramen, the branches of arteries supplying sphincters, viz., the mouth, anus, &c.

A principle of compensation also prevails throughout the arterial system, by which a less than ordinary supply from one vessel is balanced by a more than ordinary supply from another.

Section III.

Of the Venous System.

As the arteries distribute the blood from the heart to the different parts of the body, so the veins return it back again to the heart, and thus the general circulation is carried on.

As the veins have been generally described with the arteries, it will not be necessary again to enter upon their description; the following will be sufficient.

The blood from the brain is returned by the int. *jugular;* this is formed by the conflux of the sinuses at the foramen lacerum posterius, passes out through this with the eighth pair of nerves, descends in the neck, lying posterior and internal to the int. carotid art., and thence enters the carotid sheath, where it lies to the outer side of the par vagum and ext. carotid, and in front of the first stage of the subclavian art., joins the subclavian vein to form the right vena innominata.

The blood from the external parts of the face is returned by the *labial* vein, which crosses the lower jaw posterior to the labial art., descends in the neck and joins the int. jug., generally sending a branch to join the *ext. jug.;* this vein returns the blood from the temporal and int. maxillary arteries, descends in the substance of the parotid gland, becomes superficial, crosses the ext. surface of the sterno-mastoid, and at its outer margin joins the subclavian vein.

The blood from the upper extremity is returned by the *cephalic, mediam,* and *basilic* veins: the two latter join the brachial venæ comites to form the brachial vein at or a little above the elbow-joint; the cephalic runs upwards along the outer edge of the biceps nerve, then between this and the inner edge of the deltoid and the pectoralis major, and joins the axiliary vein a little below the clavicle; the *axiliary* vein conducts the blood into the subclavian, which unites with the int. jug. to form the vena innominata.

The *right vena innom.* descends almost perpendicularly into the thorax, and after a course of about an inch and an half joins the left to form the vena cava superior. The left vena innominata passes obliquely downwards and inwards, crosses the upper edge of the arch of the aorta and the vessels arising from it, and on the right side joins the right vena innom. The vena cava superior receives the vena azygos, perforates the pericardium, and empties itself into the right auricle of the heart.

The blood from the lower extremities is returned from the foot and leg by the venæ comites accompanying the arteries, which unite to form the popliteal vein. The blood from the superficial parts is returned by the int. and ext. saphenæ veins; the int. saphena is the larger, it arises about the inner ankle, ascends in the superficial fascia along the inner side of the leg immediately behind the inner edge of the tibia, crosses along the inner surface of the knee-joint, ascends along the front of the thigh, and a short distance below Poupart's lig. perforates the cribiform fascia, and joins the femoral vein; it is accompanied in the leg by the int. saphenous nerve.

The ext. saphena vein arises about the outer ankle, ascends along the back part of the leg beneath the fascia, and joins the popliteal vein; it is accompanied by the ext. saphenous nerve.

The popliteal vein terminates in the femoral, which ascends with the femoral art. and terminates in the ext. iliac vein; this joins the int. iliac to form the common iliac vein on each side; the left common iliac vein, longer than the right, passes obliquely to the right side, and unites with the right behind the iliac art. and aorta to form the vena cava inferior; the right com. iliac vein passes upwards, behind its art., to form the vena cava inferior.

The *vena cava inferior* ascends in front of the lumbar vertebræ, to the right side of the aorta, passes obliquely outwards to the right side, sinks into a deep sulcus or canal in the liver, perforates the diaphragm, receiving here the venæ cavæ hepaticæ, and almost immediately perforates the pericardium, and empties itself into the right auricle of the heart.

The vena porta returns the blood from the organs of digestion, or chylopoietic viscera, passes across the spine behind the pancreas, being here formed by the junction of the splenic and superior mesenteric trunks, enters the transverse fissure, and divides into two branches to ramify in the right and left lobes of the liver; its blood is returned by the venæ cavæ hepaticæ. The renal veins terminate in the inf. vena cava; the lumbar veins in the vena azygos.

The larger arteries are accompanied by one vein, the smaller by two, one on each side, the vena comites.

The pulmonary arteries carry black blood, the pulmonary veins red or arterial blood. In the fœtus the umbilical vein carries red or pure blood, the renal* arteries black or impure blood.

Section IV.

The Lymphatic System.

The *Lymphatics* are divided into the lymphatics properly so called, and the lacteals. The lymphatics proceed from the lower extremities, accompanying the superficial and deep veins, pass through the lymphatic glands and join the pelvic lymphatics, and then unite in front of the second or third lumbar vertebra, behind the aorta and vena cava, to form the thoracic duct. The lacteals arise from the small intestines, pass through the mesenteric glands, and join the preceding to form the thoracic duct.

The *Thoracic duct*, the trunk of the lymphatic system, passes upwards through the aortic opening of

*See Errata.

the diaphragm, between the aorta on its left and the vena azygos on its right, and enters the post. mediastinum to terminate in the left subclavian vein, as described with the post. mediastinum.

The lymphatics of the right side of the heart and neck, and right upper extremity, unite to form the right thoracic duct, which terminates in the right subclavian vein.

CHAPTER VIII.

NERVOUS SYSTEM.

The Nervous System consists of the brain and spinal cord (cerebro-spinal axis,) cerebral and spinal nerves, and the sympathetic or ganglionic system.

The brain and spinal cord are lodged in the cranium and vertebral canal, which are lined by a fibrous membrane (dura mater,) and a serous membrane (arachnoid,) for their reception. They are further covered by an immediate vesting membrane (the pia mater.)

The Spinal Cord.

When the spinal canal, and the above-mentioned coverings are laid open, the cord and the roots of the spinal nerve are seen covered by a shining investment (neurilema.) The cord extends from the pons Varolii to the second or third lumbar vertebra, where it expands, and then terminates in a lash of filaments forming the *cauda equina*. The *spinal portion* commences at the foramen magnum, the portion above that point being called *medulla oblongata*. The spinal portion does not occupy the whole of the canal formed for it by the vertebræ and dura mater, a considerable interval filled with fluid existing between them; it is, however, in some degree fixed, and its direction made to correspond to the curvatures of the spinal column, by means of the *lig. denticulatum*. This *lig.* is an exceedingly thin fold of the arachnoid membrane enclosing a few fibres, and extending the whole length of the cord; its inner edge is firmly attached to the neurilema, between the ant. and post. roots of the spinal nerves; its outer edge is serrated, the teeth being attached to the dura mater, in the intervals between the sheaths which it furnishes to the several pairs of nerves at their exit through the intervertebral foramina. The first tooth separates the vertebral artery from the hypoglossal nerve, and is attached to the occipital foramen: the last, which is the twentieth or twenty-first, is attached to the vert. opposite to the termination of the cord. The cord is not of the same size throughout, but presents three remarkable swellings, which corresponds to the great nervous plexuses. The first, *medulla oblongata*, is in the cranium; the second extends from the third cervical to the fourth dorsal vert.; and the third commences at the first lumbar vert., or a little higher up, and forms the termination of the cord. The nerves of the tongue, of respiration, and most of those of the face, arise from the first; the nerves of the super. extremity from the second; and those belonging to the infer. extremity from the third.

The cord and nerves are closely invested by a dense, fibrous, but vascular, neurilema, analagous in situation to the pia mater, but much thicker and stronger; it sends processes from its inner surface to surround the constituent fibrils, and its outer surface is covered by a mesh of vessels, which reach the cord through numerous foramina, between the fibrous filaments of which it is composed. The membrane is thrown into numerous zigzag folds, in order to allow of temporary extension produced by the various motions of the body; above, it is gradually lost in the pia mater; below, it terminates in a thin, but resisting, fibrous cord, which is attached to the dura mater lining the sacrum and coccyx; this cord, formerly considered a nerve, serves to fix the lower portion of the spinal marrow. The exact figure of the spinal cord is better seen when deprived of its neurilema. It is composed of two symmetrical lateral portions, which are separated before and behind by two fissures.

The *ant. median fissure* penetrates the third of the thickness of a cord: its floor is formed by the commissure of the cord—a simple band of homogeneous white matter, perforated by a vast number of minute vessels, extending from one half of the cord to the other. The *post. median fissure* is much narrower, and rendered almost indistinct by the neurilema; nor is it so deep as the former, excepting at the upper part of the cord; instead of a layer of white, a thin layer of cineritious matter is seen at the bottom of this fissure. On the side of the cord immediately external to the posterior roots of the spinal nerves, a line of gray matter exists, which by cautious examination, will be found to pass down to the central cineritious layer above mentioned. This *post. lateral cineritious line* subdivides the cord into an *antero-lateral*, and a *post.* tract. Immediately external to the ant. roots of the spinal nerves a faint line may also be traced; but here the substance of the cord must be broken, to arrive at the antero and lateral cinerit. line, which also reaches the central layer. If this be considered an established line of demarkation, we shall have each *antero lateral tract* anatomically subdivided into an *ant. tract* and a *lateral tract*, according to Bell and Bellingheri; the latter being included between the ant. and post. roots of the spinal nerves. The disposition of the cineritious matter, which is not the same in all parts of the cord, is seen by a transverse section. Its fundamental figure is sufficiently well represented by two crescents united by a transverse line)—(. The transverse line is the layer at the bottom of the post. fissure, and the horns correspond to the ant. and post. lateral cineritious lines.

Each half of the cord appears to be a riband-shaped layer of medullary matter, curved round the cineritious substance. This medullary riband is separable into two portions, the *antero-lateral* and *post. tracts*, which may be further divided into a series of wedge-shaped bundles of medullary fibres, which extends the whole length of the cord. The bases of the wedges are turned outwards and the apices inwards; they are united by their sides, and so give the medullary riband, which they form, the proper incurvation. The apices of these wedges are not ranged in a uniform curved line; the consequence of this irregularity is a denticulated appearance of the outer surface of the cineritious layer.

Medulla Oblongata, or Bulb.

The medulla oblongata is that portion of the cord extending from the margin of the foramen magnum to the pons Varolii. In shape it resembles a quadrilateral pyramid, with its base above and its apex below. The ant. surface presents a median fissure, not so deep as that in the cord; below it is interrupted by the decussation of the ant. pyramids; above it terminates at the pons, in a *foramen cæcum*. The *ant. tracts* of the cord here presents two bodies, which are called, from their shape and position,

The anterior Corp. Pyramidalia.

These bodies are as long as the medulla, and lie on each side of the ant. fissure; they commence below by their narrow extremities, which is about a line and a half in thickness, and ascend, gradually diverging to the pons, where they are twice that in size; they then suddenly diminish to pass above that substance. The c. pyram. are composed of white fasciculi, and appear, although considered by Rolando as indistinct superadded bodies, to be the ant. tracts of the cord, augmented by fresh nervous matter.

The decussation of the ant. Pyramids.

This remarkable interchange of filaments of the anterior pyramids takes place about the lines below the pons, it is formed by three or four large fasciculi, which regularly and successively intercross. The decussation comprehends the ant. two-thirds of the medulla, and also takes place from before, backwards, as the ant. portion of the ant. tract, or pyramid of one side, may be traced backwards to be continuous with the more lateral fasciculi of the other. No other decussation exists in the medulla.

Corpora Olivaria.

These two bodies are situated external and somewhat posterior to the ant. pyramids. They are two white oval eminences about six lines long, embedded to half their diameter in the medulla; in fact, they are inserted between the anterior and post. tracts, and are only separated from each other at the median line by antero-poster. transverse filaments, the continuations of which (*fila. arciformia*) bind them, as it were, in their situation: they contain a layer of gray matter, crumpled into a kind of concentric frill.

Corp. Restiformia; post. pyramids; infer. Cerebellar Peduncles.

The posterior tracts of the cord bear these titles; they ascend diverging on the post. surface of the medulla, till they reach the cerebellum. The post. fissure is converted into a triangular depression (*fourth ventricle*) by their divergence. This ventricle, like the post. fissure, is lined by cineritious matter, marked by a few medullary striæ, so as to bear altogether a vague resemblance to a pen (hence *calamus scriptorious;*) a median furrow represents the stem, the medullary striæ the feather, and the commencement of the divergence the point.

Filamenta Arciformia,

a term given by Santorini to some medullary filaments leading from the anterior to the posterior tracts; they are important, because they show that the infer. cerebellar peduncles are not formed solely of the post. tracts. The antero-post. fibres, before mentioned, wind horizontally round the ant. pyramids, and pass, one set of fibres immediately below the corp. olivaria, and another on the inner side of those bodies, to the corp. restiformia. Similar, but deeper seated, fibres are described by Mr. Solly, to cross and obliterate the post. cineritious line, and then join the restiform bodies.

Fasciculi Innominati.

When the ant. and post. pyramids are removed, a dense grayish white substance is seen to occupy each half of the cord; like the olives, they are only separated by the ant. post. fibres. This substance, which is continuous with the corresponding lateral tract, is prolonged to the optic thalami, &c.

Isthmus.

The medulla oblongata is connected to the cerebrum and cerebellum by certain prolongations which occupy an intermediate situation between them. The isthmus comprises the *crura*, or *pedunc.* cerebri, the contin. of the *fascic. innom.* the *super.* and *middle peduncles* of the cerebellum, *pons Varolii, tubercula quadrigemina,* and *valve of Vieussens.*

The *pons Varolii* is a mass of medullary matter of a square shape, resting upon the basilar process of the occip. bone, and supporting the anterior pyramids in their passage to become the crura cerebri. Its ant. and post. edges are free, but latterally it ascends on each side of the crura to constitute the middle peduncles of the cerebellum. A section of the pons shows it to be composed of antero-posterior and transverse medullary fibres, intermixed with gray matter, which gives each slice a striated appearance; the former fibres belong to the crura, the latter to the middle peduncles. The inferior surface is marked by a groove, which is caused by the projection of the two corp. pyramidalia; it generally lodges the basil. art.

The *peduncles of the cerebrum,* or *crura cerebri,* are

two large white round cords about six inches long, which continue the ant. pyramids in the cerebrum. They diverge in proceeding to that organ, leaving a triangular space, which anter. is occupied by the corp. mammil. and tuber. cinereum, and poster. by the inner portion of the fascic. innom. Each crus is often intersected by two perpendicular white tracts, one from the testes and valve of Vieussens, the other from the inner surface of the crura. The crura are separated from the fascic. innom. by a stratum of dark gray matter (*locus niger;*) their size is strictly relative to that of the corresponding hemispheres.

Peduncles of the cerebellum. The *inferior* are the two corp. restiform, strengthened by the fili. arciform. the *middle* are constituted by the prolongations of the pons, which might be called their commissure; the *superior, pedunc. processus cerebelad. testes; intercerebral commissure,* are two lamellæ which arise in the substance of the lateral lobes of the cerebellum, on each side of the median line, and extend upwards and forwards, apparently to terminate in the testes. They form the ceiling of the aqueduct of Sylvius, and their inner edges, which are proximate, are united by the valve of Vieussens.

Valve of Vieussens. A thin semi-transparent lamella, which unites the adjoining edges of the sup. cerebellar peduncles. It also covers the aqueduct of Sylvius.

Tubercula Quadrigemina. These are four roundish eminences, placed two on each side of the median line, behind the thalami; the anterior two are called *nates,* the posterior two *testes.* The former are oblong, and of a grayish colour, the latter are rounder, and more inclined to white. In animals they constitute the optic tubercles, and are in some joined into a single pair; they are single in the fœtus, and in the adult the separation is not complete. The nates are united to the thalami, and the testes to the intercerebral commissure. On the outer side of the thalami, a little anterior and below the tubercula quadrigemina, are two little swellings called *corpus geniculatum, externum* and *internum*, the lateral tracts strengthened by fascicul. innom. and a fasciculus from each c. oliv. terminate in these tubercles and thalami—the whole may therefore be considered as one connected system.

Cerebellum.

The cerebellum is placed in the inferior occipital fossæ, beneath the tentorium, which separates it from the posterior lobes of the brain. It is elliptical in shape, and composed of two symmetrical halves, or lobes, united at their fore part by an intervening median lobe. The lateral lobes are separated behind, and below, by a median fissure, which receives the falx cerebelli; below and in front, the fissure dilates into a *notch*, which partly encloses the back of the medulla oblongata. When the lateral lobes are pulled apart, the median lobe is seen continuous on each side with the former, and its two halves united by a prominent median raphè.

The *Super. Vermiform process* is this raphè on the upper surface of the lobe; it is prominent in front, and covers the valve of Vieussens; behind it is almost lost, where it joins the infer. vermiform process, or under part of the raphè, on the back of the lobe. The *Infer. Vermiform P.* is marked by a few rings, it is seen in the middle of the notch, and forms the post. paries of the fourth ventricle, into which project several unimportant eminences, designated by a few meaningless terms.

The Fourth Ventricle.—This is a rhomboid shaped cavity, extending from the aqueduct of Sylvius, by which it communicates with the third ventricle to the point of the calamus, where it presents an *orifice* leading into the sub-arachnoid cellular tissue. It is bounded below and in front by the fascia innom.; laterally by the peduncles of the cerebellum, above, by the sup. peduncles of the cerebellum, valve of Vieussens, median lobe, and inf. vermiform process, and neurilema of the cord. The *Aqueduct of Sylvius* runs beneath the valve and super. pedunc. to the third ventricle. The *inf. orifice* is rendered interesting by the recent researches of Magendie, on the seat of the cerebro-spinal fluid.

The superficies of the lobes of the cerebellum is grooved by numerous horizontal curved lines, which divide the organ into a number of segments and lamellæ, analogous to the convolutions of the cerebrum, laid one on the other like the leaves of a book. The continuity of these lamellæ is not interrupted at the median lobe, but are merely drawn out of their direction as they pass from one lateral lobe to the other; the antero-post. vertical section of the median lobe shows its *arbor vitæ*. This appearance is produced by two medullary trunks, super. and inf. springing from a central medullary nucleus, and dividing into many orders of branches. The primary branches, six in number, correspond to so many segments of the organ; each segment divides into smaller segments, lamellæ, and lamellules; the latter are extremely irregular. A stratum of cineritious substance invests these divisions, and dips in the intervals between them. Similar sections of the lateral lobes display an arbor vitæ, with branches of the same character to each of them. By attentive examination, the nuclei and their medullary ramifications, are discovered to be composed of medullary plates, two of which, at least, enter into the formation of the smallest lamellule. In the centre of each of the lateral nuclei is a small round body, *corpus rhomboideum*, or ganglion of the cerebellum, resembling the c. olivaria, particularly in the arrangement of the stratum of gray matter which it contains. The peduncles of the cerebellum, three on each side, enter into, or proceed from the lateral nuclei.

Cerebrum.

The cerebrum is that part of the nervous mass which fills the entire cranium, excepting the inferior occipital fossæ. The greater part of the brain is composed of convolutions, which spring as it were from the isthmus, situated on the basilar process and body of the sphenoid, and extend in every direction till they entirely fill every corner of the cranium. The *superior surface* of the brain is divided into two symmetrical lateral halves, or *hemispheres*, by a deep median fissure, which receives the *falx major*. The whole of this surface, and those of the hemisphere next the falx, are marked by longitudinal elevations and depressions (*Circumvolutions and sulci,*) which have their corresponding depressions, *digital fossæ*, and elevations on the interior of the skull. The greater part of the *infer. surface* is also formed by the convolutions of the hemispheres, which are here subdivided into three lobes: the *ant. lobes* rest upon the orbitar processes of the frontal and smaller wings of the sphenoid bone; the *middle lobes* are lodged in the sphenoidal fossæ, the *post. lobes* on the tentorium. The ant. are separated from the middle by a deep fissure (Sylvii;) but the line of demarkation between the middle and the posterior lobes is merely a faint line, produced by the upper edge of the petrous bone. The anterior and posterior lobes are only separated by the ant. and posterior continuations of the great median fissure; but the middle lobes have between them, resting upon the basilar process and body of the sphenoid, the isthmus and several remarkable eminences, seen on the base of the brain: viz.

Optic Tracts.—Two fasciculi of white matter, which arise on each side of the corp. genic. ext., wind round each crus at the point where it terminates in the corpus striatum, and then run inwards and forwards to form the *commissure* of the optic nerves on the olivary process of the sphenoid bone. These fasciculi, with the crura cerebri, intercept a lozenge-shaped space, which contains the fascic. innominati, c. mammillaria, tuber. cinereum, infundibulum, and pituitary gland.

Interpeduncular space; or space left by the divergence of the crura, is occupied behind by the lower surface of the fasciculi innom., in front by the c. mam.

Corpora Mammillaria.—Two small round white eminences, situated between the crura, their ant. surface imbedded in the tuber. cinereum, their posterior in the fascic. innom. Although tolerably distinct below, they are united above by a soft gray substance, which forms part of the floor of the third ventricle. They are composed of a central portion of gray, covered by a cortex of white matter, apparently derived from the anterior pillars of the fornix.

Tuber cinereum.—A soft gray substance which fills the interval between the corpora mammil. and the optic. fascic. and their commissure. It is a continuation of the soft substance which unites the c. mam. and with it forms part of the floor of the third ventricle.

Infundibulum.—A reddish cord, about two inches in length, directed downwards and forwards from the tuber cinereum, to the ant. lobe of the pituitary gland, to which its apex is attached. It contains a funnel-shaped canal, which above commun. with the third ventricle, but ceases at the gland. It is composed of a cylinder of fibrous pia mater, lined by the same soft gray substance as is seen in the third ventricle.

Pituitary Gland.—A small body of grayish white substance, weighing from five to ten grains; it is situated in the sella turcica of the sphenoid bone, where it is confined by the circular and cavernous sinus, and a fold of dura mater. It is composed of an anterior larger, and post. smaller lobe, separated by a fibrous membrane. They both contain numerous small vessels.

The commissure of the optic nerves being removed, the anterior part of the floor of the third ventricle is exposed; it is formed by that soft gray substance which is a continuation of the tuber cinereum, it is covered by the neurilema of the optic nerves, which seem to derive a few filaments from it. A little more in front is the ant. reflected extremity of the corp. callosum, joining the anterior lobes of the brain; still more in front the ant. and inf. portion of the great median fissure.

Behind the pons Varolii is the post. and inner. ext. of the great median fissure, which when widened, shows the post. extremity or bulb of the c. callosum, uniting the post. lobes. Between the bulb and the upper surface of the c. quadrigemina is a fissure, where the pia mater, as the *tela choroides*, enters the ventricles. This is the fancied situation of Bichat's celebrated arachnoid foramen, which does not exist (Cruveilhier.) Here also, surrounded by the tela choroides, is the pineal gland. The pia mater also enters the ventricles by another fissure, (*Grande fente celebrale*, Bichat,) which extends on each side, from the extremity of the former to the fissure of Sylvius. It enters the lateral ventricle between the hemisphere and outer sides of the thalamus. It only admits the pia mater.

Fissura Sylvii.—The fissure corresponds to the wing of Ingrassias, and commences at the ant. end of the *grande fente*, extending outwards, so as to form a deep interval between the ant. and middle lobes. It contains the middle arteries of the cerebrum. The angle of the middle lobe, formed by the junction of the two above-mentioned fissures is called *subs. perforata:* it is white, and pierced by numerous small vessels.

Corpus callosum, (grand transverse commissure:) (exposed by a horizontal section on a plain a little above it, or, by separating the hemispheres;) it is a stratum of medullary matter, about three and a half inches long, uniting the two hemispheres, at the bottom of the median fissure. The *upper surface*, convex, presents a median furrow produced by two parallel longitudinal strata of white matter, formed by the arteries resting here; under these are perceptible numerous transverse fibres, of which this commissure is principally composed. The *infer. surface*, concave, covers the lateral ventricles, and also presents longitudinal fasciculi; anteriorly, in the median line, it receives the *septum lucidum*, sent up from the fornix; posteriorly it joins the latter body, and terminates just above the *c. quad.* in a bulbous extremity, which unites the post. lobes: anteriorly it is reflected over the c. striata and terminates beneath them just in front of the third ventricle, where it unites the ant. lobes. Two white fasciculi (*peduncles of the c. cal.*) run, parallel to the optic fasciculi, from this extremity to the subs. perforata before mentioned.

Fornix; longitudinal commissure.—A triangular medullary arch in the lateral ventricles, resting upon the thalami. It is composed of two lateral fasciculi joined in the median line. The apex of the fornix terminates in two fasciculi, (*ant. pillars,*) or continuations of the lateral fasciculi, which arch over the fore part of each thalamus as the corpus callosum did over the c. striata ; descend between those two bodies, through the gray matter on the sides of the lateral ventricles and behind the ant. commissure, to the c. mammillaria, to form the medullary cortex ; the post. angles are continued as the *post. pillars*, outwards and downwards in the lateral ventricles, to form the corp. fimbriata. The base is attached to the bulb of the corp.

callosum. Between the upper surface of the thalamus and ant. pillar on each side, a space (*for. Monro*) is left, by which each lateral communicates with the third ventricle.

Septum Lucidum.—The triangular interval between the fornix and corp. callosum is closed by two delicate semi-transparent lamellæ, which are sent up from the contiguous edge of each half of the fornix, to the under and fore part of the corpus callosum. The interval beween them is called the fifth ventricle, which sometimes contains a few globules of fluid: it does not communicate with the other ventricles. The septum lucidum separates the lateral ventricles.

Lateral Ventricles.—Exposed by dividing the corp. callosum on each side of the septum lucidum. They commence in each anterior lobe a little before the beginning of the third ventricle, forming the anterior cornu, run upwards and backwards, and a little beyond the termination of the third vent., turn round the post. part of the thalamus, beneath which they are reflected, to terminate in the middle lobe, close to the fissure of Sylvius, only a little below and behind where they commenced, forming the inferior cornua. At the point of reflection it sends a prolongation (digital cavity) into the posterior lobe, forming the posterior cornua.

The superior, or straight portion of the lat. ventricle.—The floor of this portion is formed by the upper surfaces of the corresponding corp., striatum, and optic thalamus.

Corpus Striatum.—A grayish, pyriform body deeply embedded in the fore part of the floor of the lateral ventricle. Its largest extremity lies in front of the thalamus, but behind it runs tapering along the outer side of that body to the point where the lateral ventricle begins its reflection. Internally it corresponds to the gray matter which forms the sides of the ant.

part of the third ventricle, and to the outer side of the thalamus. Below, it sinks deeply into the post. and inner part of the anterior lobe. Its interior presents a remarkably striated appearance, produced by the diverging fibres of the cerebrum, of which it is supposed to be the great ganglion.

Thalamus nervi optici.—An oblong brownish-white body, situated in the floor of the lateral ventricle, behind and internal to the c. striatum. Its ant. extremity is separated from the c. striatum by the ant. pillars of the fornix, and its outer side from that body by the *tenia semicircularis*. Its inner surface forms part of the side of the lateral ventricle. It is covered by the tela choroides and inf. surface of the corresponding lateral stratum of the fornix; its interior is composed of gray matter.

The *tenia semicircularis* is a narrow white band, separating the corp. striat. from the thal.: it is covered by a vein which is enclosed in a fold of the lining membrane of the ventricle.

In *the inf.* or *reflected portion of the lat. ventricle, is* 1, *hippocampus major*; *cornu ammonis*—a portion of a convolution projecting from the floor; 2, *corpus fimbriatum*, a narrow dense band, a prolongation of the post. pillar, or angle of the fornix which runs on the hippocampus.

In the post. portion, or *digital cavity*, is seen the *hippocampus minor*, also a projecting convolution. It is connected to the h. major by a narrow band which is also connected with the fornix.

Third Ventricle.

This is merely a fissure separating the optic thalami. Behind and above, its sides are formed by these bodies, but before and below it is lined by a peculiar soft gray matter, which is continuous with the tuber cinereum. The floor, concave, is constituted posteriorly (between the crura,) by the fascic. innom.; in the middle by the tuber cinereum, c. mammil. and infundib.; anteriorly by the reflected portion of the c. callosum. Its sides are united by three commissures—viz. 1, *com. mollis*, a soft gray substance like the tuber ciner., situated just in front of the thalami; 2, *com. ant.* a white cord just in front of the ant. pillars of the fornix, uniting the thalami; 3, *com. post.* a similar cord uniting those bodies just in front of the t. quadrigemina.

The *aqueductus Sylvii* is a canal leading under the tuber quadri. from the third to the fourth ventricle. The third ventricle contains, 1, the opening of this aqueduct; 2, that of the infundibulum; and 3, 4, the foramen, or foramina of Monro.

Pineal Gland.—A small conical body resting on the fissure separating the t. quadri., just behind the post. commissure. Its base is attached to the thalami, by two peduncles, which are also attached to a white stratum (*Pineal comm.,*) which extends between the thalami, above the post. commissure. It frequently contains irregular concretions of phos. lime and animal matter (Pfaff,) and is closely invested by the tela choroides.

Membranes of the Brain.

The surfaces of the brain are supplied with blood by a vascular membranous network, (pia mater,) which dips between all the convolutions, and enters the ventricles, where it forms thick vascular meshes, (*plexus choroides,*) which also receives additional arterial branches from the base of the brain. The *tela choroides* spreads over the third ventricle and thalami, but under the fornix, the pia mater enters into post. fissure to form it: it dips into the third ventricle to form its plexus choroides, and enters the lat. ventricles through the foramen of Monro on each side to form a choroid plexus for each of them. The pia mater also enters the lat. vent. through the *grande fente*. The plexus of the third passes through the aqued. of Sylvius, to form the plexus ch. of the fourth ventricle.

Dura Mater.

This is a strong fibrous membrane, which lines the skull as a periosteum to the bones, and sends septa to separate and support the different divisions of the nervous mass. The construction of these partitions is best understood by supposing a fold of membrane to be pinched up on the line occupied by these septa, and drawn, more or less, into the interior of the cavity. The *falx major* separates the two hemispheres; it commences by its point being attached to the crista galli of the ethmoid bone, and its base is continuous on each side with the upper lamina of the tentorium. Its upper, or cranial edge, convex, is attached along the inner mesial line of the cranium, and contains the *super. long. sinus;* its lower edge, concave, is free, and contains the *inf. long. sinus.* The *tentorium* separates the cerebrum from the cerebellum; its circumference, or cranial edge, corresponds to the lateral occipital grooves, upper edge of the petrous bone, and it term. at the post. clinoid processes; its inner crescentic edge surrounds the medulla oblongata, its horns being prolonged, under the terminations of the outer edge, to the ant. clinoid processes. The *falx cerebelli* separates the lateral lobes of the cerbellum: its upper portion, or base, is continuous with the inf. lamina of the tent.; it terminates by a pointed end at the foramen magnum.

The dura mater consists of two membranes, the inner one forming the septa, the outer one remaining attached to the bone; it is lined internally by a reflexion of the arachnoid membrane, externally it is closely connected to the inner surface of the bones of cranium by numerous vessels.

The *sinuses*, fifteen in number, are composed of the inner membrane of a vein enclosed between these two layers.

1, The *super. longitudinal sinus*, corresponds to the

cranial edge of the falx: it contains the cordæ Willisii, which are said to preserve its venous orifices patent; 2, *infer. longitudinal sinus*, inf. edge of the falx; 3, 4, *lateral sinuses*, cranial edge of the tentorium; 5, *straight sinus*, contin. with the inf. long. sinus, runs in the diamond shaped canal, between the conjoined laminæ of the falx major, minor, and tentorium; 6, 7, *cavernous sinuses*, on each side of the sella turcica; traversed by the third, fourth, first branch of the fifth and sixth pair of nerves; the carotid artery, and carotid plexus of the sympathetic, all of which are separated from the blood by the venous membrane; 8, 9, *inf. petrosal sinus*, inf. or occipital angle of the petrous bone; 10, *transverse sinus* joins the last two across the basilar process; 11, 12, *super. petrosal sinuses*, upper angle of the petrous bone; 13, *circular sinus*, surrounds the pituitary gland, communicating on each side with the cavernous sinuses; 14, 15, *occipital sinuses*, occipital foramen.

The *straight sinus* receives the inf. long., s. vena Galena, and the inf. and middle cerebral and cerebellar veins. The *petrous sinuses* receive the cavernous, circular, and transverse sinuses. The *lateral sinuses* receive, 1, the *super. long. s.*, *straight s.*, and *two occipital sinuses*, opposite the occipital tuberosity, the conflux being termed *torcular Herophili;* and 2, the *petrosal sinus* at some point of their course to terminate in the int. jugular veins.

The dura mater descends through the foramen magnum, to which it closely adheres in the vertebral canal as a sheath, which encloses the spinal cord, and forms a distinct sheath for each pair of spinal nerves. It is loosely attached by means of distinct ligamentous slips to the post. comm. lig., but a peculiar reddish semi-fluid fat, traversed by many vessels, separates it from the arches and lig. subflava; below it forms a large sac, which surrounds the chorda equina, and is distended by the cerebro-spinal fluid. A large space exists between the sides of the cord and sheath also generally filled with this fluid.

The *arachnoid* is a very fine serous sac placed between the dura mater and pia mater. It covers the convolutions, and is also carried into the third ventricle by the venæ Galeni, through the post. fissure, it then passes into the other ventricles and infundib. through its four openings. A similar sac is found between the spinal cord and its sheath of dura mater. The arachnoid passes from one convolution to the other without entering the sulci; it is also stretched from the tuber. cin. to the pons, and from the lobes of the cerebellum to the cord, large corresponding spaces are thus left between it and the pia mater, with which the numerous minor spaces between the convolutions and ventricles communicate. The whole of the spaces are filled with the cerebro-spinal fluid.

The arachnoid membrane presents a similar arrangement in the spinal canal investing the inner surface of the dura mater, and the outer surface of the pia mater, and forms the lig. denticulata.

The *pia mater* is the immediate investing membrane of the cord, it is much more dense but less vascular than that of the brain.

Structure of the Brain.

Each hemisphere, by transverse sections, is seen to be formed by a large central nucleus of medullary matter united by means of the corpus callosum. Each nucleus divides into three segments, which subdivide into the convolutions. The convolutions, unlike the lamellæ of the cerebellum, are only a few lines (from nine to fourteen) deep, and meander in various directions. In size and sit. a few of them only are constant, most of them differ greatly in different brains. Their use is to augment the extent of surface of medullary matter, and thereby obtain a more extensive stratum of that peculiar gray matter which is supposed to play such an important part in all nervous phenomena.

LATERAL TRACTS.

Since the beautiful theories of Gall and Spurzheim, which have given a peculiar tone and direction to all subsequent researches respecting the construction of the nervous centres, anatomists are accustomed, in describing the connections and relations of the different parts of the nervous system, to employ such terms as origin, divergence, expansion, &c., terms which must be understood only as expressive of the peculiar views of the various authors who make use of them. The amount of our present real knowledge of the construction of the nervous system may be stated as follows:—

1. The *anterior tracts* of the cord decussate, by the principal part of their filaments, at the commencement of the medulla oblongata, and may then be traced upwards, first as the anterior pyramids, and then as the crura cerebri, above the pons Varolii, forming the diverging fibres of the cerebrum, to the corpora striata; further, their white medullary filaments may be traced through the gray substance of which these bodies are composed, and to make use of the figurative, from the side opposite to that at which they entered, they are seen to diverge forwards, upwards, and backwards, so as to compose much of the medullary centres of the hemispheres, finally terminating by being lost in the cortical layer of gray matter.

2. The *lateral* tracts of the cord becoming *fasciculi innom.*, in the medulla oblongata, and strengthened by two fasciculi from the olivary bodies, ascend parallel to and upon the upper part of the crura (the separation between them being marked by the locus niger) to be continuous without any line of demarkation whatever with the thalami: they are also connected to the tuber quadri. From the opposite side of each thalamus white diverging fibres are seen to arise, and diverge in all directions, intermixing with the posterior fibres from the corpora striata, like them also to terminate in the cortical gray layers

of the convolutions. Some fibres from the thalami traverse the corp. striata.

3. The *posterior tracts* are continued in the medulla oblongata, as the restiform bodies, or inf. cerebellar peduncles; these may be still further traced, without any decussation, to the central nuclei of the lateral lobes of the cerebellum, where they pass through the corpus dentatum or ganglion of the cerebellum, and thence diverge to form its hemispheres. Previous to their entrance into the cerebellum it will be recollected that they are joined by the filamenta arciformia.

4. The convoluted portion of the *hemispheres* is thus composed of the medullary radiations from the central nuclei, or corp. striata and thalami, covered by a layer of gray matter.

5. The cerebellum is composed as before stated.

The nervous system is made up of two symmetrical lateral halves, united by certain medullary strata called *commissures;* thus,

6. The lateral sides of the cord are united by a uniform layer of *soft white matter*, seen at the bottom of the ant. median fissure.

7. The lateral lobes of the *cerebellum* are united, 1, by the middle lobe, the raphe of which, seen as the two vermiform processes, indicates that it also formerly consisted of two lateral halves; 2, by the pons Varolii, which is continued upwards round each crus cerebri, as the middle peduncles of the cerebellum, to the central medullary nucleus of each lateral lobe of that body; 3, by the valve of Vieussens.

8. The central nuclei (c. striata et thalami) of the hemispheres are united by, 1, the ant. commissure; 2, the posterior commissure; and 3, the Pineal commissure.

9. The two hemispheres are united by the great transverse commissure, or corpus callosum.

10. Each hemisphere is united in its different parts; 1, by the inferior longitudinal commissure or fornix,

which by its ant. pillars before, and its post. pillars behind, is connected to each lobe; 2, by the superior longitudinal commissure, composed of longitudinal fibres, which run in the substance of the hemispheres in the same curved lines as (but a little above) the corpus callosum, and

11. The cerebrum and cerebellum are united by the intercerebral commissure, or superior peduncles of the cerebellum.

The commissures are formed by the *transverse or converging* fibres, which arise in the external cineritious substance, and pass inwards to the opposite hemisphere.

The hemispheres appear to be composed of the medullary striæ from the corpora striata; of similar striæ from the thalami; and striæ which are continuous on each side with the transverse fibres of the corpus callosum, all intermixing and radiating in various directions till they arrive at the surface of the convolutions; the latter are formed by an outer stratum of gray matter, and an inner medullary stratum thrown into a series of folds (convolutions;) each convolution being composed of a rind of this gray and white matter, containing the terminal ends of those numerous white striæ from the corp. striata, thalami, and c. cal.

Comparative anatomy will, in some degree, assist in rendering our knowledge of the brain a little more clear. It is seen, for instance, that the brains of some animals, especially fishes, are composed of two linear series of lobes, which answer to the olfactory lobes, corpora striata, thalami, tubercula quadrigemina and middle lobe of the cerebellum, in man. In animals of a grade still lower than those referred to, the brain is composed of little nodules, of less number and size, so that it is difficult in some cases to distinguish which should properly be called thalamus, &c.; but enough is learned by such a comparative consideration of these primitive strata, or accumulations of nervous matter, to show the probable use of the hemispheres and the parts relating to them. The human brain, in fact, consists of certain primitive nuclei, covered by superadditions of nervous matter in the shape of the two hemispheres. If the organ be traced through the phases which it presents at successive periods of fœtal existence, the primitive nuclei are observed to become gradually covered by two thin membranes which seem to spring from their sides; they represent the hemispheres: it is not until the hemispheres have nearly finished their progress that the convolutions make their appearance.

The corpus callosum, fornix, septum lucidum, the lateral ventricles, and the parts they contain, as the hippocampi, corpus fimbriatum, are dependents of the hemispheres, and therefore are never found when the latter are absent.

The uses of the following parts are still enveloped in mystery:—

1. *Corpora albicantia*, absent in birds and reptiles, very large in fishes; supposed to be, like the hippocampi, a convolution in a different form.

2. *Pituitary gland* and *infundibulum*.

3. *Pineal gland*, or conarium; the seat of the soul (Descartes.) A plug to obstruct the orifice leading from the third to the fourth ventricle (Magendie.)

4. The relative uses of the cineritious and medullary structures. Until a period subsequent to birth they are indistinguishable from each other; they appear to be formed simultaneously, and only acquire their distinctive characters at some distant period. It would not be unprofitable to take a general view of this remarkable gray matter in its various relations to the medullary substance, but we have only space to observe, that it is found principally at the central terminations of nerves, and other medullary media, which may thus be considered as mere internuncii.

The brain is supplied with blood by the int. carotid and vertebral arteries; the former enter, one at each side, through the carotid foramen, in the petrous portion of the temporal bone, pass through the canal, and ascend by the side of the sella turcica; they here divide into the anterior and middle cerebral posterior communicating arteries. The anterior passes forwards and inwards, is joined to its fellow by the ant. communicating branch, winds round the ant. extremity of the corpus callosum, and terminates in the callos. art. The middle passes upwards and outwards into the bottom of the fissure of Sylvius, and supplies the ant. and mid. lobes. The posterior communicating branch passes backwards to join the posterior cerebral art., and forms the side of the circle of Willis.

The vertebral arteries enter by the foramen magnum, ascend forwards and inwards, and unite in front of the pons Varolii to form the basillar art. This ascends along the groove in the pons, and at its upper margin divides into the superior art. of the cerebellum and post. art. of the cerebrum, to their destination in the cerebellum and post. lobe of the brain; the latter joins the post. communicating branch, as just described, and thus completes the circle of Willis. Each vertebral art. gives off, before forming the basilar, two branches, the anterior and posterior spinal arteries; these descend along the corresponding surfaces of the spinal cord, anastomosing with similar branches from the deep cervical, intercostal and lumbar arteries, and the inferior artery of the cerebellum to the under surface of this part. The blood from the brain is collected in the sinuses, and is returned by the int. jug. vein; the veins from the spinal marrow form numerous venous plexuses around the transverse processes of the vertebræ, and empty themselves into the adjoining venous trunks.

DIVIDED INTO CEREBRAL, SPINAL, AND SYMPATHETIC.

Spinal Nerves.

It has long been considered that the experiments of Sir C. Bell and Magendie have left no doubt, that, of the double roots by which each pair of spinal nerves is attached to the cord, sensation depends on the posterior or ganglionic root, and motion on the anterior; but it is by no means ascertained that they derive such powers from corresponding tracts of the cord. The cord is stated to be composed of an anterior or motor portion, and a posterior or sensory portion, which are joined together *somewhere* in the lateral tract; now the one or other power may be fairly assigned to that point where the anterior and posterior roots are respectively attached; but the precise origin of them is by no means determined. Cruveilhier denies the existence of different orders of nerves; he could not even perceive any difference between the result from a division of the anterior, and that from the division of the posterior roots.

The spinal nerves are thirty-one in number, including the sub-occipital. There are eight cervical, twelve dorsal, five lumbar, and six sacral.

Origin of the Spinal Nerves.

The spinal nerves arise by two linear series of filaments from the anterior and posterior cineritious lines and are, therefore, separated by the lateral tracts, and the ligamentum denticulatum. In the adult they cannot be traced into the anterior of the cord, so as to ascertain their precise points of origin, but, through the semitransparent cord of a fœtus seven or eight months old, numerous delicate nervous filaments, which compose the anterior and posterior spinal nerves, are seen to traverse the central gray matter, and may be followed even to the posterior tracts.

Immediately after the exit from the cord, the filaments divide into sets of eight or ten, which converge so as to form one of the roots of the spinal nerves. The anterior and posterior roots soon approximate, but do not unite, and each is enclosed in a distinct sheath of dura mater.

Differences between the two Roots.

The *anterior* roots are smaller, excepting the first, or sub-occipital, whose anterior root is the larger, near the median line, and their filaments are not attached to the cord exactly in the same line. They have no ganglions.

The *posterior* roots are larger, their filaments are attached in succession to the posterior cineritious line, and each presents, when arrived at the corresponding intervertebral foramen, a characteristic swelling or ganglion, which is situated within the foramen.

A junction of the two roots now takes place; the anterior root, however, is not entirely free from the ganglion, which frequently receives several filaments from it.

Branches of the Spinal Nerves.

The conjoined cord of the two roots divides into, 1, a middle or ganglionic branch; 2, a posterior branch; and, 3, an anterior branch.

The ganglionic branches belong to the sympathetic nerve.

The posterior branches, the smaller, excepting the first two, supply the parts on the back of the trunk.

The anterior branches supply the fore part of the trunk, and the extremities.

Post. Branches of the Cervical Nerves.

The first escapes between the occipital bone and

atlas, on the inner side of the vertebral artery; it lies in the triangle formed by the rect. post. major, and obliqui, which it supplies.

The second, the largest, escapes between the arches of the first and second vertebræ, beneath the lower edge of the inf. obliquus, and ascends under the cervical muscles, which it pierces close to the occipital artery, to ramify with it on the post. part of the cranium. *Ram.* to the adjoining muscles.

The third, arrived between the two complexi muscles, divides into an occipital, and transverse or cervical branch.

The fourth, fifth, sixth, seventh, and eighth, are much smaller than the former, and after supplying the cervical muscles, terminate in the integuments. The first, second, and third, also form a species of plexus, beneath the complexi muscles.

Posterior Branches of the Dorsal Lumbar Nerves.

The eight superior branches divide into *external*, or *muscular branches*, which subdivide in the space between the sacro-lumbalis and longissimus dorsi muscles, to supply those m., &c., and *musculo cutaneous branches*, which, after a curious winding course, pierce the latissimus dorsi near the spine, and run horizontally outwards to the skin.

The four last dorsal, and the five lumbar, pierce the common mass of the long dorsal muscles, supply them, and then wind horizontally round the loins, some under the latissimus dorsi, and some between the layers of abdominal muscles, which they pierce here and there, to terminate on the skin of the abdomen.

The *posterior branches of the sacral nerves* pierce the muscular mass, on the back of the sacrum, to ramify in the skin over the sacrum and nates.

Anterior Branches of the Spinal Nerves.

CERVICAL PLEXUS; formed by the ant. branches of the first (not sub-occipital,) second, third, and fourth cervical nerves. The first and second at their exit from the vertebral foramina anas. by an ascending and descending branch, so as to form a curious loop around the transverse process of the atlas; from this loop branches supply the recti muscles and *anastomose* with the first cervical ganglion and the spinal accessory. The cervical plexus is situated under the post. edge of the sterno-mastoid muscle, between the rectus anticus and ins. of the splenius colli; it is covered by adipose tissue and lymphatic glands. *Ram.*

Ascending set. 1, *Superficial branches;* two or three ascend over the sterno-mastoid to the integuments; *ram. superficial. colli,* to the integuments over the ear and parotid gland; a filament anastomoses with the seventh pair. 2, *Deep branches;* three or four small nerves to the sterno-mastoid, digastric, splenius, &c.

Descending set. 1, *Internal branches;* a. an *anastomosing branch,* to join the *descendens noni;* b, *phrenic nerve;* composed of branches of the second, third, and fourth cervical nerves; it descends diagonally across the ant. scalenus, enters the chest between the subclavian artery and vein, descends between the pericardium and pleura accompanied by an artery, (the left nerve winding around the apex of the heart,) to the diaphragm, on which it ramifies: it communicates with the last cervical ganglion. 2, *External branches:* a. *supra clavicular;* b. *supra acromial;* and c. *sternal,* to the integuments covering those bones; *external respiratory* (Bell,) one of the several deeper-seated descending branches, which descends on the side of the thorax to the serratus magnus muscle.

Posterior set. a. A branch to anastomose with the

spinal accessory; b. branches to the trapezius, levator scapulæ, and rhomboidei.

BRACHIAL PLEXUS: composed of the anterior branches of the four last cervical and the first dorsal of the spinal nerves; these branches descend parallel with each other, between the scaleni muscles, and only constitute a plexus in the middle of the axilia behind the tendon of the pectoralis minor.

Collateral branches, 4, a. *infra clavicular*, branches to the subclavius, levator scapulæ, and rhomboidei; b. *supra scapular nerve*, passes through the supra scapular notch, to the supra and infra spinati muscles; c. *sub-scapular nerves*, two or three branches, which sometimes come from the circumflex to the sub-scapularis m. d. *thoracicæ*, four or five in number, descend behind and before the clavicle, to the surface of the pectoral muscles.

Terminal branches, 6 in number.

1. *Circumflex*, turns backwards under the lower edge of the sub-scapular muscle, and winds round the surgical neck of the humerus with the inferior circumflex art., to terminate in the deltoid; it supplies the teres major et minor, latissimus dorsi, and skin over the shoulder.

2. *Internal cutaneous*, descends on the inner side of the arm, pierces on the fascia a little above the elbow, divides into filaments, which generally accompany the veins, and ramify in the skin nearly as far as the wrist.

3. *External cutaneous*. (perforans casserii,) pierces the coraco-brachialis m., descends on the outer side of the arm, pierces the fascia a little above the elbow, descends beneath the skin, to which it is distributed by separate filaments as far as the anterior and posterior part of the carpus; it also supplies the muscles along which it passes; hence its name.

4. *Ulnar nerve*, descends on the short head of the triceps, passes between the internal condyle and ole-

cranon, through the space between the two heads of the flexor carpi ulnaris, soon gains the inner side of the ulnar artery, which it attends to the hand: it terminates in, a, *a superficial branch*, which supplies the two sides of the little, and the ulnar side of the ring finger; and, b, a *deep branch*, which forms an arch by inclining outwards, beneath the flexor tendons, so as to supply the interossei, and terminate in the adductor pollicis; c, *dorsalis carpi ulnaris*, turns over the ulna beneath the flexor ulnaris, to supply both *dorsal* edges of the little, and the ulnar edge of the ring finger.

5. *Median nerve*, (the largest branch,) arises by two heads, which embrace the axillary artery: it descends 1st, on the outer side of the brachial artery, along the outer edge of the coraco-brachialis muscle; 2d, it crosses the artery to lie on its *inner* side, resting on the brachialis anticus; 3d, it passes between the two heads of the pronator teres, and descends between the flexor sublimis and the flex. profund., between the tendons of which two muscles it enters the palm of the hand, where it divides into five digital branches. *a*, The three first correspond to the two edges of the thumb and radial edge of the fore finger; *b*, the *two* last supply both edges of the index, and middle, and the radial edge of the ring finger: *c*, *interosseous anticus;* attends the anterior interosseous artery, with which it is lost on the back of the carpus; *d*, *palmar cutaneous*, to the skin on the palm of the hand.

6. *Musculo spiral:* runs with the superior profunda artery in the spiral groove of the humerus, and lies with that vessel between the brachialis anticus and supinator longus, where it divides into: a, *r. radialis*, soon gains the *outer* side of the radial artery, along which it descends to within a short distance of the wrist; it now turns over the outer edge of the radius, its terminal branches supplying the two dorsal edges of the thumb, index, and middle fingers, and the radial

edge of the ring. (The dorsal and palmar edges of the little, and the ulnar edge of the ring finger, are thus supplied by the ulnar nerve; the dorsal and palmar edges of the thumb and other fingers, with the radial edge of the ring finger, by the median and radial branch of the musculo spiral,) b, *ramus profundus* (post. interosseous,) pierces the supinator brevis, descends with the *post.* interosseus artery, and supplies the two layers of extensor muscles. Each principal artery of the arm has thus its attendant nerve.

The skin of the upper arm is supplied, 1, by branches from the three upper intercostals, (*cutaneous of Wrisberg*,) which traverses the axilla; and, 2, by branches from the *circumflex* and *spiral* nerves.

Anterior branches of the dorsal Nerves.

A, INTERCOSTAL NERVES, 12 in no., extend from the first dorsal intervertebral foramen to that between the last dorsal and first lumbar. As a general rule, they gain the corresponding intercostal spaces, where they run (*below* the art.,) between the intercostal muscles, and divide, about midway between the spine and sternum, into external, or *cutaneous* branches, and proper *intercostal* branches or continuations of the nerve: the latter pierce the thorax, near the sternum, to supply the inner part of the pectoral muscles; the former having pierced the chest at the point abovementioned, send *anterior cutaneous* branches, which run a regular parallel course, and post. *cutaneous* branches, which run over the ext. surface of the latissimus dorsi. The intercostal nerves communicate close to the vertebræ by ascending and descending branches, and send one or two filaments to the corresponding sympathetic ganglia.

The first dorsal nerve belongs to the brachial plexus, and sends therefore but a small intercostal branch. The second intercostal is very tortuous before it reaches the space; the cutaneous branches of the second and third are the *cutaneous of Wrisberg*. The fourth, fifth, sixth, and seventh dorsal intercostals follow the general rule; but the eighth, ninth, tenth, and eleventh nerves traverse the costal attachments of the diaphragm to terminate on the abdomen, by cutaneous and muscular filaments, which, in ramifying on the abdominal muscles, precisely follow the above rule. The twelfth intercostal sends a muscular branch to the abdominal muscles as above, but its cutaneous branch is large, and descends nearly to the side of the ilium; it also sends a branch, which, with the ant. branches of the four first lumbar nerves, constitute the lumbar plexus.

THE LUMBAR PLEXUS, besides the above nerves, receives a branch from the last lumbar, to make up the uninterrupted chain of connection which includes all the spinal nerves; it is situated on the second, third, and fourth lumbar transverse processes, covered by, or even partially buried in, the psoas muscle.

Collateral branches, 4 in No.

1. *Sup. abdominal branch (ilio-scrotal)* arises from the first lumbar nerve, perforates the psoas, crosses the quadratus, runs round the abdomen parallel to the last dorsal branch (which it much resembles,) and at the iliac crista gets between the trans. and int. oblique m., where it divides into an *abdominal* and *pubic* branch; the former is distributed to the muscles like the last dorsal; the latter, sometimes joined by the next branch, runs along the cord with which it escapes to terminate in the skin of the labia or scrotum.

2. *Inf. abdominal branch (ileo-scrotal minor)* crosses the quadratus and iliacus m., running parallel to, but below, the former, which it sometimes joins; when it does not, it is seen as a second branch, making its way through some part of the inguinal region, to the same parts as the former.

3. *Inguino-cutaneous* (external cutaneous.) Arises

from the second and third lumbar nerve, traverses the psoas, crosses the iliacus below the last nerve, and escapes between the two ant. iliac spinous processes. *Ram. a*, anterior or *femoral* pierces the fascia, sends a cutaneous branch to the ant. and outer parts of the thigh down to the knee; *b*, *posterior* to the skin, over the tensor vaginæ, glutei, &c.

4. *Genito-crural* (external pudic.) Arises from the second lumbar nerve, traverses the psoas muscle, descends on its ant. surface, with the ext. iliac art., and just before it reaches the groin divides into two branches, *a*, *internal* or *scrotal* branch, crosses the artery, enters the inner ring, runs beneath the cord, and terminates, like the ilio-scrotal, in the labia or scrotum; *b*, a *cutaneous* branch, which terminates on the ant. and inner part of the thigh.

Terminal Branches 2 in No.

1. *Obturator;* arises from the third and fourth lumbar nerves, runs beneath the inner margin of the psoas m. along the brim of the pelvis, escapes with the obturator artery through the obt. foramen, and in a space between the obturator ext. and pectineus m. divides into an *anterior* branch, which goes to the add. brevis et longus and gracilis, and a *posterior* branch to the obt. ext., and. brevis, and magnus.

2. *Ant. crural nerves;* arises from the second, third, and fourth lumbar nerves, descends along the outer edge of the psoas m., between it and the iliacus internus, and immediately below Poupart's ligament divides into 1, a *superficial set* of branches, usually two in no., which supply the skin on the fore and inner part of the thigh, as far as the knee; 2, a *deep set*, of these, a branch (*n. saphenus major*) runs in the sheath of the femoral vessels, becomes subcutaneous just below the knee, and accompanies the int. saphena vein to ramify on the inner part of the dorsum of the foot, the *saph. min.* runs along the sartorius. Numerous branches,

divided into external and internal, supplies the muscles on the ant. and outer part of the thigh.

SACRAL PLEXUS, the communicating branch of the last lumbar, and the ant. branches of the four first sacral nerves, unite into a large flat cord, which is situated deeply in the posterior part of the pelvis, on the pyriformis muscle: it is covered by the int. iliac artery, and (on the left side) by the rectus.

Collateral Branches, 4 in No..

1. *Hæmorrhoidal,* lower part of the rectum, sphincter, and mucous membrane.
2. *Vaginal, uterine, and vesical* filaments, to those organs.
3. *Pudic nerve,* it exactly follows the course of the int. pudic artery; Ram. *a,* a *branch,* which pierces the great sacro sciatic ligament, and winds round the tuber ischii to be distributed by branches corresponding to the inf. hæmor. and superf. perineal art., to the skin about the perineum and scrotum; *b, ramus dorsalis penis,* to the skin of the penis in the male, clitoris in the female.
4. *Smaller sciatic nerve,* it escapes from the pelvis at the lower edge of the pyriformis m. *Ram. a, muscular,* distributed in all directions to the gluteus maximus; *b, cutaneous,* to the skin over the flexor muscles as far as the knee, by distinct *post.* and *int.* cutaneous br.

Terminal Branch.

Great sciatic nerve; escapes at the lower border of the pyriformis muscle, and descends between the tuber ischii and great trochanter to the middle of the thigh, where it terminates in two branches, the *post. tibial* and *peroneal m.;* it rests upon the external rotator and adduct. m., covered by the gluteus maximus and hamstring muscles. Ram. *a,* branches to the obturator int., gemelli, gluteus max., ext. rotator, and adductor muscles; *b*, a posterior, middle, and external cutaneous branch.

Peroneal nerve (external terminal branch,) descends along the tendon of the biceps muscle, winds round the neck of the fibula, between that tendon and the external head of the gastrocnemius, where it is subcutaneous, and between that bone and peroneus longus m. it divides into three *terminal branches*.

1. *Two recurrent branches*, which supply the upper part of the extensor muscles, like the tibial recurrent artery.

2. *Musculo-cutaneous*, descends in the course of the peroneal trunk, in the substance of the peroneus longus m., pierces the fascia some way above the ankle-joint, and divides into two branches, one, (internal) ramifies, by two digital branches, on the inner border of the foot, and first and second toes; the other (external,) by means of three digital branches, which correspond to the digital spaces, ramifies on the outer border of the foot, and dorsal surfaces of the three last toes.

3. *Ant. tibial*, runs beneath the upper part of the common extensor m. to the interosseous ligament, descends along the ant. surface of the corresponding artery, runs beneath the annular ligament of the ankle, in the sheath of the ext. prop. pollicis, and divides into a *branch*, (deep digital,) which runs under the dorsal artery to supply the adjoining edges of the first and second toes; *and one*, which, after sending delicate filaments to the interosseous spaces, terminates in the extensor brevis digitorum m.

The *collateral branches* of the peroneal nerve, are. 1, *communicans peronei*, which arises in the popliteal space, pierces the fascia, and unites with the communicans tibiæ, to form the ext. saphenus nerve, under which name it descends with the vein to the outer side of the ankle, and sometimes to the outer side of the foot; its branches are calcanian and malleolar; 2, *a cutaneous* branch to the skin over the outer and back part of the fibula,

Post. tibial nerve, (intern. terminal branch,) descends, as the *popliteal* nerve on the outer back part of the pop. vein, and, as the post. tibial nerve, along the back part of that artery, to the internal annular ligament, under which they are both included in one fibrous sheath. *Collateral branches.*

1, *Muscular;* 2, *communicans tibiæ*, unites with the communic. peronei as before described; 3, *branches* to the gastrocnemii and soleus.

The *terminal branches* are, 1, *internal plantar nerve*, (the larger,) enters the sole of the foot, close to the calcaneum and above the flexor muscles, and gives off four digital branches, which supply, as the median nerve did in the hand, both edges of the three first toes, and the inner edge of the fourth; 2, *External plantar nerve*, (smaller,) runs outwards between the flexor accessorius and flexor brevis; it gives off, *a*, a *deep branch*, which enters the sole of the foot to supply some of the muscles; and *b*, *three branches*, one of which runs to the outer edge of the foot, the other two (*digital*) supplying both the edges of the little toe, and the outer edge of the fourth toe. A small branch also anas. with the last digital branch of the internal plantar n.

The fifth and sixth sacral nerves, which do not enter into the formation of the sacral plexus, are extremely small; the latter is merely a delicate filament scarcely observable. The fifth anas. above, with the fourth, and, below, with the sixth; the latter pierces the sacro-sciatic ligament to terminate by three cutaneous branches upon the coccyx.

CEREBRAL NERVES, 9 PAIRS.

1. OLFACTORY, *Or.* by numerous filam. from the under surface of the olf. lobes; *term.* roof of the nares septum and two sup. turbin. bones. The lobes are

attached to the brain by a pedicle, which sends, 1, a *middle gray root* through the subs. perf. to the *soft* com.; 2, an ext. long root, to the post. edge of fiss. Sylvii: 3, an int. short root, to the inner part of the ant. lobe.

2, OPTIC, *Or.* optic commiss. *term.* retina. To form the comm., 1, the inner fibres of the opt. fascic. meet transversely; 2, the mid. fib. decussate; 3, the outer fibres proceed without decussating.

3. MOTORES OCULI, *Or.* by a linear series of striæ, from the adjoining edges of the fascic. innom. at the ant. edge of the pons; *term.* in the orbit by 1, *ram. super.* to lev. palp. and sup. rectus; 2, *ram. inf.* (larger,) to rect. inf. et int. inf. obliq.,—radix ext. to ophth. gang.

4. PATHETICI (delicate long.) *Or.* valve of Vieussens; *term.* sub, obliq. joined to the opthal. nerve.

5. TRIGEMINI, (spinal nerve, double) both roots are attached to the side of the pons; *traced*, the sensy. to post. half of the medulla obl.; the motor to ant. pyram. within the pons. The semilunar sensy. gang. (Gasserian,) rests on the petrous bone, and gives off, 1, ram. opthalm.; 2, ram. sup. max.; 3, R, inf. max. OPHTHALMIC. *Ram.* a. *lacrymal;* 1, fil. malar; 2, fil. palp., to palp. and conjunc.; 3, fil. lacrymal, to the gland, *anas.* with the fascial and s. max. n. b. *frontal;* fil. 1, supra orbitar; 2, *supratrochlear.* c. *nasal;* 1, long root of lent. gang. 2, fil. *infratrochlear.;* 3, *ethm. ant.;* enters the skull, and descends through the crib. plate, sends a fil. to the septum, and one to the skin of ala nasi: 4, two fil. ciliar. SUP. MAXILLARY. *ram.* a, *orbitar,* 1, fil. lacrymal, 2. temp. malar; b, *post. dental,* 1, *fil.* post. et ant., to post. teeth, gums, and buccinator; c, *ant. dental,* antrum, and ant. teeth; d, *infra* orbitar (terminal.) INF. MAXILLARY, joined by the motor root *ram.* a, *muscular,* (motor root,) *fils.* deep temp., masseteric, buccal, pterygoid; b. *auricular,* (superf. temp.) *anas.* with the *fascial;* c, *gustatory or*

lingual, to papil. edge and tip of tongue, *anas.* with 1, chorda tymp., 2, infra dental; d, *infra dental*, fils., mylo-hyoid, mental.

6. ABDUCENTES, *Or.* ant. pyramids at the post. edge of the pons; *term.* ext. rectus m.; *anas.* with *ophthalmic* n. and 1st *cerv. gang.*

7. { AUDITORY, *portio mollis*, } attached. *Or.* fossa
   { FACIAL, *portio dura* } between the corp. restif. and pons; *traced* the former round the c. restif. to the striæ of the calamus in the 4th vent., the latter, through those bodies to the stem. PORTIO MOLLIS; *term.* by, a, *ram.* to the cochlea, b, *ram.* to the vestibule and semicirc. canals. *Anas.* with *port. dura.* PORTIO DURA. *Ram.* In the fal. canal; 1, receives the Vidian, 2, gives off *chorda tympani*, which crosses the tymp. and joins the gust. nerve; 3, receives a filam, through the bone, from the vagus (Arnold.) Out of the ear. *Ram. collateral;* 1, *post. auric*, 2, *styloid*, 3, *submastoid. Ram. term.;* 1, *cervico-facial* r. buccal, mental, cervical; 2, *temp. facial; anas.* with *fifth pair;* r. temp. orbitar, and infra arbitar.

8. { GLOSSO PHARYNGEAL, } *Or.* The two former
   { PNEUMOGASTRIC, } by a linear series of
   { SPINAL ACCESSORY, } fil. from the corp. restiform.; the latter, also by a series of fil., from the lat. tract of the cord, extending in a line behind the lig. dentic. from the vagus to the fifth cervical nerve. The three escape by the foramen lacerum, the former in a fibr. canal distinct from that of the two latter. GLOSSO PHARYNGEAL. In the canal it presents *a ganglion*, the br. of which are; a, *fil.* (*Jacobson*) penet. the inner wall of the tymp., and gives, 1, fil. to the carot. plexus, 2, fil. to anas. with Vidian, 3, fil. to the *otic.* gang.; b, *anas. fil.* to facial nerve. In the neck; a, *anas. fils.*, to vagus, and spinal access.; b, r, *digast.* et *stylo-hyoid;* c, *pharyngeal*, to the plexus; d, *tonsillary;* e. *Lingual* (term.,) papil., and muc. mem. at the root of the tongue. VAGUS. In its canal; a

ganglion, *ram.* a, *fil.* penet. the jugul. foss. to anas. with the facial in its acqueduct, it first sends a fil. to r. Jacobson. At its exit; *anas, fil.*, to spinal access., hypoglossal, pharyngeal, and sup. cerv. gang. In the neck; a, *r. pharyn.* to plexus; b, *sup. laryngeal*, to epiglottis, crico-thyroid muscle; c, *r. to join the cardiac n.* In the thorax: the right vagus enters between the subc. art. and vein; the left, behind the subc. vein, in a triang. interv. between the com. carot. and sub. art. *Ram.* a, *n. recurrens*, the left round the aortic arch, the right round sub. art.; 1, *fils. to join the cardiac nerves and plexus*, 2, *fil.* œsoph., and tracheal, 3, pharyngeal, 4, term. *laryngeal muscles*, and muc. memb.; b, *ant. pulm. brs.;* and c, *post, pulm. brs.* to the pulmon. plexuses; d, *œsophageal plexus*, surround the organ; e, *gastric plexus*, the left vagus ant., the right poster., the latter joins the solar plexus. SPINAL ACCESSORY; sometimes forms the post. root of the first cerv. pair. In the canal, it is joined to the vagus. At its exit; a, *anas. fil.* to glos.-pharyn., vagus, and three upper cerv. nerves; b, *r. sterno mastoid*, which m. it perf. to term. in the trapezius.

9. HYPOGLOSSAL; *Or.* by a linear series of fil. like the ant. roots of the spinal nerves, between the c. oliv. et ant. pyram.: it trav. the ant. condyl. for. *Ram. collat.;* a, *anas fils.*, to vagus, sub. cerv. gang., first and second cerv. nerves, and ling. nerve; *Ram. term.;* a, r. to the glossal muscles; b, *r. descend. noni*, anas. with cerv. plexus, and sends fil. to the infra hyoid muscles.

SYMPATHETIC SYSTEM.

The muscular tissue of the heart and digestive canal, the genital and urinary passages, the arteries, especially those of the trunk, are supplied by this system; brs. are also sent to the organs of sense. The veins, lymphatic vessels, and glands, and serous membranes, receive none from it. The system consists of, 1, the ganglia of the sympathetic; 2, detached ganglia in the head, chest, and abdomen; 3, nervous plexuses, derived from them. The ganglia, although they derive their energy solely from the cerebro-spinal system, are, to a certain extent, independent sources of nervous power, which, by means of their plexuses, is developed in such combinations as are suited to the associated functions they are designed to regulate. The SYMPATHETIC NERVE consists of three cervical, twelve dorsal, four lumbar, and four sacral ganglia, extending in a chain from the first cerv. to the last sac. vert. and connected by a nervous cord. They commun. by 1, *Ext. br.*, one in the cervical, two in the other regions with the several spinal pairs; 2, by *Ascend.* and *Descend. br.* with the gang. next above and below; 3, they send *Intern. br.* principally to attend the arteries; and 4, a *Visceral*, or *splanchnic* set. ARTERIAL BR., a, *Nervi molles;* from the first cerv. gang., to accomp. the ext. carot., and its br. forming plexus, named from them, as thyroid, lingual, meningeal, &c.; b, *Nerv. vertebralis*, from the last cerv. gang. accomp. the vert. art. and its br. (The int. carot. and br. are supplied by the cavern. plexus;) c. *fil.* to the subt. are. and its prim. br. and to the brach. plex.; d, *Aortic fil.* from the dorsal gang. to accomp. each interc. art.; e, *Lumbar Aortic br* . The remaining vessels in the neck, thorax and abdomen, are attended by fil. from the vagus, cardiac gang. and semil. gang. VISCERAL. FIL. a, *Pharyngeal*, to the plexus; b, *Laryngeal*, joins the sup. laryng. n.; c, *Cardiac nerves*, three on the right, sometimes but two on the left; arise from the three cerv. ganglia, and descend in the thorax, where they are joined by card. br. of the vagus and the recur. n.; *term.* in the card. gang. and plexus; d, *Splanchnic nerves*, two in number; the sup. from sixth, seventh, eighth, and tenth thoracic gang.; the inf. from the eleventh and twelfth gang. The former joins the semilunar gang., the latter the renal plexus; e, *splanchnic br.* from the lumbar gang. to the lumbar aortic, and inf. mesent. plexus.

CEREBRAL GANGLIA, a, *Lentic. Gang. sit.* between the ext. rectus and opt. n.; *Post. br.;* 1, fil. or long root, from the nasal n.; 2, a fil. short root, from inf. br. of the third pr.; 3, fil. to the caver. plex. *Ant. br.;* sup. and inf. ciliary fil. (two cil. br. come from the nasal.) b, *Meckel's Gang.; sit.* pteryg.-max fissure; *Post. br.* 1, Vidian nerve; *Int. br.* 2; naso-pal. n.; *Descending br.* 3, post. pal. n.; 4, *fil.* to carot. plex.; 5, fil. to super. max. n. c, *Otic Gang.* (Arnold;) *sit.* inner side of inf. max. n. immediately below the for. oval. *Ram.* 1, long root, which is the middle fil. or Jacobson's n.; 2, short root, to the inf. max. n.; 3, fil. to lax. tymp. m.; 4, fil. to auric. temp. n.

CARDIAC GANG. and PLEXUS; *sit.* under the aortic arch, and between it and the trachea. *Comp.* cardiac nerves and card. br. of the vagus and recurrt. *Ram.;* fil. to form the ant. post. coron. plex.

SEMILUNAR GANG. and *solar plexus; sit.* on the cœliac axis: the ganglion on each side rec. the grt. splan. n.; the plexus *rec.* br. from the sm. splan. n., and right vagus; sometimes the right phrenic.

The br. of the solar plexus accompany the neighbouring art. and branches, forming plexuses, which are named from them: the hepatic, splenic, and gastric plex., commun. with the vagus. The *Sup. mesenteric plex.* also receives post. fil. from the lumbar gang. The *Renal plex.* rec. the lesser splanch. nerves, and forms the spermatic, or ovarian plexus. The *Inf. mesent. plex.* rec. fil. from the lumbar gang. and the hæmorr. plex. &c. The *Hypogastric plexus* is sit. on each side of the rectum; it is *comp.* of lumbar and sacral fil., and supplies the rectum, bladder, uterus, vagina and testicles.

The *Pharyngeal plexus* is sit. on the side of the pharynx; it is comp. of fil. from the vagus, glosso-ph. and sympathetic. The *Pulmon. plexus*, ant. (smaller) and post. (larger,) are *comp.* principally of the vagus; it rec. br. from the sympath.; *ram.* to the bronchial tubes. *Œsophageal plexus*, comp. principally of the right and left vagus.

The *Sup. cerv. gang.* commun. with the cereb. nerves; 1, by a *carot. fil.*, which forms the carot. plexus, and joins the cavernous plexus; it receives a fil. from the Vidian and Jacobson's n., and sends fil. to the sixth pr. of n. (The *cavern. plex.* rec. fil. from the third, fifth, and sixth prs. and Lent. gang.; sends fil. along the br. of the int. carotid.) 2, by fil., which commun. at the base of the skull, with the glosso-phar. vagus, hypoglossal, and facial nerves. The *last sacral gang.* is situated on the front of the lower extremity of the sacrum (*impar.*) branches from it descend to the anus.

Although the sympathetic nerve is here described as proceeding from the spinal nerves, it may be more proper to regard it as a distinct nervous system, which communicates by filaments with the spinal and cerebral nerves. Its filaments preside over the functions of organic life, and act independently of the will.

19

CHAPTER IX.

OF THE BONES, LIGAMENTS, AND JOINTS.

The Bones of the Head

are divided into those of the *cranium and face*.

The *cranial bones* are eight in number, viz., one frontal, two parietal, one occipital, two temporal, one sphenoid, and one ethmoid.

FRONTAL.—*Sit.* upper and ant. part of skull. *Ant. surface* convex, is slightly depressed on the mesial line, indicating the union of the two pieces of which this bone consists in the young subject, on each side of this bone is convex, and covered superiorly by the scalp, inferiorly it is irregularly concave, and immediately beneath again convex, opposite the frontal sinus, laterally is a concavity forming part of the temporal fossæ, bounded above by the temporal ridge. *Post. surface*, concave, lodges the ant. lobes of the brain; on the mesial line is a projecting spine, which divides as it passes upwards into two, leaving a depression: the former gives attachment to the falx cerebri, the latter lodges the commencement of the sup. longitudinal sinus.

Inferior surface presents the orbital plate, this is concave beneath and forms the roof of the orbit in front, and to the outer side is a deep depression for the lachrymal gland, and internally a slight depression for the pulley of the sup. oblique musc. irregularly convex above, it supports the ant. lobes of the brain.

The *circumference* of the frontal bone presents a slight projection in the centre; from this, passing outwards, it forms the coronal suture with the parietal bone; beneath this, it articulates with the ant. inf. angle of the parietal, and anteriorly terminates in the ext. angular process which artic. with the malar bone, passing *inwards* from this arc, the superciliary arch, notched near its inner ext. by the *superciliary notch*, sometimes a foramen, the *int. angular process* for artic. with the os nasi, and on the mesial line the *nasal spine*, to support in front the nasal bones and posteriorly artic. with the nasal lamella of the ethmoid; the *posterior edge* of the orbital plate artic. posteriorly with the lesser wing of the sphenoid, internally with the orbital plate of the ethmoid, and more anteriorly with the os unguis. The foramina are nine, viz., with the ethmoid one foramen cæcum, and four, two, anterior, and two posterior int. orbital; two superciliary or supra-orbital, two frontal sinuses. It articulates with four bones of the cranium; two parietal, sphenoid, and ethmoid, and eight bones of the face, nasal, sup. maxil. lachrymal, and malar. The *frontal sinuses* exist in the anterior and inner part, and extend backwards into the orbital plates.

PARIETAL.—*Sit.* Superior and lateral part of cranium. *External surface* convex, presents in centre parietal eminence; below this a rough surface, a little flattened to form part of the temporal fossæ above, and the squamous sutures with the temporal bone beneath. *Inner surface*, irregularly concave, is marked by the convolutions of the brain, and anteriorly by the branches of the mid. meningeal art. *Circumference.* The *upper* edge forms with the opposite bone the sagittal suture, it is grooved internally for the sup. longitudinal sinus; the *anterior* forms with the frontal the coronal suture; the posterior joins the occipital to form the lambdoid suture; the inferior, arched, is overlapped by the squamous portion of the temp. bone forming the squamous suture. The *ant. inf. angle*, prolonged downwards to artic. with the sphenoid, is grooved internally by the middle art. of the dura mater; the *post. inf. angle* obtuse, joins the mastoid part of the temp., and is grooved by the lateral sinus.

The parietal artic. with five bones, frontal, sphenoid, temporal, occipital, and the opposite parietal.

OCCIPITAL.—*Sit.* post. inf. part of cranium. *Ext. and post. surface* irregularly convex, smooth above, is covered by the scalp; it is separated from the inferior portion by the central protuberance, to which is attached the lig. nuchæ, and on each side by the superior transverse ridge, for the attachment of the occipito-frontalis and trapezius muscle; beneath this, the bone forms a part of the base of the skull, it presents a rough surface for the insertion of the complexus and splenius capitis, then the *inferior transverse ridge* and another rough surface for the insertion of the recti postici and sup. oblique; on the mesial line is a rough spine for the attachment of the lig. nuchæ; more in front is the *foramen magnum*, oval in shape; it transmits the spinal marrow and its membranes, vertebral arteries, and spinal accessory nerves; in front of this is the *basilar process*, which passes upwards and forwards, and joins the body of the sphenoid; its superior surface, concave, receives the pons Varolii; the inferior, slightly convex, gives insertion to the recti cap. antic. musc. posteriorly, and anteriorly forms the roof of the pharynx. On the under and lateral ant. part of the foramen magnum are the *condyles*, convex, and placed obliquely to artic. with the atlas; in front of each is the ant. condyloid foramen, for the exit of the ninth nerve; and behind, the post. condyloid foramen for entrance of a small vein; to the outer side is the *jugular process*, bounding the foramen lacerum posticus, and giving insertion to the rectus cap. lateralis.

Internal surface, irregularly concave, is divided into four fossæ by a crucial ridge; the superior fossæ lodge the post. lobes of the cerebrum; and the inferior, the lateral lobes of the cerebellum; the transverse portion of the ridge gives attachment to the tentorium, and is grooved for the lateral sinuses; the superior portion of the vertical ridge gives attachment to the falx cerebri, the inf. to the falx cerebelli; in the centre is the int. occip. protuberance corresponding to the torcular Herophili, in front is the inner opening of the foramen magnum, smooth and larger than the external.

Circumference.—The superior pointed extremity, and from this, outwards to a projecting spine, articulates with the parietal, forming the lambdoid suture, in front of this it joins the temporal, and still more anteriorly the body of the sphenoid. This bone artic. with six bones—two parietal, two temp., sphenoid and atlas. Its foramina are the ant. and post. condyloid, and the foramina lacera postica, in basi cranii; these latter are completed by the petrous bone, and transmit the int. jug. vein, the eighth pair of nerves, and some small meningeal arteries.

TEMPORAL.—*Sit.* lateral, middle, and inferior part of skull, irregular in shape, each is divided into three portions; the squamous, mastoid, and petrous. The *squamous* is thin and overlaps the parietal. *Ext. surface,* convex behind, concave in front, forms part of the temp. fossæ, from near its interior part the *zygomatic process* runs forwards, and terminates in a suture, which joins the malar, to complete the zygoma, this presents behind its roots the meatus auditorius externus, and on its lower margin a projection for the attachment of the ext. lat. lig. of the lower jaw; internal to this is the *glenoid cavity*, divided into two by the Glasserian fissure, which transmits the chorda tympani, and the tendon of the levator tympani; behind the fissure, the cavity contains a portion of the parotid gland; in front is the articular concave surface for the condyle of the lower jaw, in front of this is the transverse root of the zygoma convex to complete the temporo-maxillary artic.

The squamous plate is marked internally by the convolutions of the brain and small meningeal vessels.

The mastoid portion occupies the post. part of the temp.; *ext. surface* convex, presents the *mastoid process* for the insertion of the sterno and trachelo-mastoideus, *internally* it is grooved by the lateral sinus, its interior is cellular. Internal to the mastoid process, and in the base of the skull, is the digastric groove for the origin of the digastric musc.

The *petrous portion* runs forwards and inwards from the preceding portions; triangular in shape, its base is attached to them, its apex is free. *Inferior surface*, in front of the foramen lacerum posterius is the aqueduct of the cochlea, a small foramen leading into the cochlea; external to this is the styloid process for the attachment of the styloid muscles and ligaments; post. and ext. to this is the *stylo-mastoid* foramen for the exit of the portio dura; in front and int. to the styloid process is the *carotid foramen*, leading to the carotid canal, and in front of this is a rough surface, to which are attached the muscles of the velum and palate.

Superior surface, irregular, forms part of the middle fossa of the cranium; in front is a groove for the Casserian ganglion; behind this a slit-like aperture, the *hiatus Fallopii*, through which the Vidian nerve passes to the aqueduct of Fallopius, and a convex surface formed by the superior semicircular canal. This surface is separated from the posterior by a thin edge, grooved by the sup. petrous sinus, and giving attachment to the tent. cerebelli. *Posterior surface*, nearly flat, presents the meatus auditorius internus, for the exit of the seventh pair of nerves; behind this is a slit-like aperture, which leads to the aqueduct of the vestibule. The petrous bone contains the internal ear.

Circumference.—The squamous plate articulates in front with the sphenoid; above it overlaps the parietal, and thus forms the squamous suture; behind this, the mastoid portion articulates with the post. inf. angle of the parietal, then with the occipital; in front of this the petrous bone articulates with the side of the basilar process, forming the foram. lacer. posterius, in front with the body of the sphenoid, forming the foram. lac. anticus, and then with the spine of this bone; the temp. articulates with five bones, the parietal, sphenoid, occipital, malar, and inf. maxillary.

SPHENOID. *Sit.* In the centre, base and lateral portions of the skull. It is divided into a body, wings, and processes.

The *body* is in the centre, from it anteriorly projects on the mesial line the *azygos process* to artic. with the nasal lamella of the ethmoid above, and the vomer beneath, on each side of which is the opening into the sphenoidal sinus, partially closed by the bones of Bertin, on its *superior surface* is the *stella turcica*, a deep fossa for the reception of the pituitary gland, bounded in front by the ant., and behind by the post. clinoid processes, for the attachment of the edges of the tentorium cerebelli. In front is the *olivary process* for the lodgement of the commissure of the optic nerve, behind is the *clivus* which joins the basilar process. At each side of the olivary process is the *optic foramen* for the exit of the optic nerve and ophthalmic art.; and still more exteriorly, a thin plate of bone, the *lesser wing of the sphenoid*, or *wing of Ingrassius*, which artic. by its ant. edge with the orbital plate of the frontal.

The *great wing* proceeds from the side of the body outwards, forwards, and upwards, its *anterior* surface forms the *orbital plate*, and is found in the outer wall of the orbit, the internal concave forms part of the middle fossa of the cranium, the *external* concave is divided into two by a crest, the superior portion forms part of the temporal fossa, the inferior part of the zygomatic fossa. From the back of the great wing proceeds the *spinous process*, which is received into a retreating angle between the squamous and petrous bones; in it is the *spinous foramen* for the passage of the mid. art. of the dura mater, and from it descends the *styloid process*, to which is attached the int. lat. lig. of the lower jaw; ant. to this is the *foramen ovale* for the exit of the inf. maxillary nerve, and still more anteriorly, the *foramen rotundum* for the exit of the sup. maxillary nerve. Between the wings is the *foramen lacerum orbitale*, sometimes closed exteriorly by the frontal, which transmits the third, fourth, first division of the fifth and sixth nerves, some sympathetic filaments, one head of the ext. rectus, and the ophthalmic vein. From the junction between the body and great wing descends the *pterygoid process*, this soon divides into two plates, the *int. and ext. pterygoid*, separated by a deep fossa posteriorly, but united anteriorly; the int. plate bounds the posterior nares, on its outer side, and terminates inferiorly in a hook-like process, round which the tendon of the tensor palati plays; the ext. plate gives attachment on its outer surface to the ext. pterygoid muscle, and on its innner to the int. pterygoid; at its root is the fossa navicularis for the origin of the tensor palati, and above this the Vidian canal: inferiorly these plates leave an angle into which the palate bone is received.

The sphenoid artic. with all the bones of the cranium, and five of the face, two malar, two palate and the vomer. The processes are twenty-six, four clinoid, one olivary, one ethmoidal, four wings, two spongy, or bones of Bertin, two temporal, two orbital, two spinous, two styloid, two hamular, and four pterygoid. The foramina are fourteen proper and eight common. The *proper* are, two optic, two lacerated orbitals, two round, two oval, two spinous, two Vidian, and two sinuses. The *common* are two foramina lacera-antica, two spheno-maxillary fissures, two spheno-palatine and two posterior palatine canals.

The *bones of Bertin* or *spongy plates of the sphenoid*, are two triangular plates, which assist in closing up the sphenoid sinuses in front.

ETHMOID. *Sit.* In the anterior part of the cranium, between the orbits, and forming the roof of the nasal cavities; of a cuboid shape, it consists of a transverse and vertical plate and two lateral masses; *superior surface*, on this is seen the *transverse* or *cribriform plate*, perforated by numerous foramina for the exit of the olfactory nerves, two of these nearest the centre transmit the nasal nerve; in the centre is the *crista galli*, for the attachment of the falx cerebri; this is continued downwards forming the *vertical plate*, the lower portion, the larger, is the *nasal lamella;* it forms part of the septum narium, and articulates with the nasal spine of the frontal, nasal bones, triangular cartilage, vomer and sphenoid bone: the *lateral masses* are composed of delicate cells in the centre, the spongy bones internally, and the orbital plates externally: the *spongy bones*, the *superior* and *middle*, are separated by the superior meatus; of the cells, the posterior are the smaller, and open into the superior meatus, and anterior open into the middle meatus. The orbital plate is quadrilateral, and forms part of the inner wall of the orbit, its upper edge artic. with the frontal, and here forms the internal orbital foramina, the lower edge artic. with the sub. maxillary and palate, in front with the os unguis, and behind with the sphenoid. The ethmoid artic. with two bones of the cranium, and frontal and sphenoid, and eleven of the face, the nasal, superior maxillary os unguis, palate, inferior spongy, and vomer.

The principal *sutures* uniting the bones of the head are the *coronal, sagittal, lambdoidal, squamous,* and *addita omentum, suturæ lambdoidalis,* and *squamosæ.*

THE BONES OF THE FACE

are fourteen in number—viz., two malar, two sub. maxillary, two lachrymal, two nasal, two palatine, two spongy, one vomer, and one inferior maxillary.

MALAR. *Sit.* upper and outer part of cheek. Irregularly quadrilateral. *Ext. surface*, convex, presents a few small holes for the passage of vessels and nerves; *sup.* it is prolonged upwards to join the frontal, and form the outer edge of the orbit; *internally* it joins the superior maxillary; between these it forms part of the lower margin of the orbit; behind, the malar is concave, and forms part of the temporal fossa, and *superiorly*, concave, to form the orbital plate. The lower edge of the malar terminates in the zygomatic process, which joins that from the temporal to form the zygoma; from it arises the masseter muscle.

SUPER. MAXILLARY. *Sit.* central part of the face. Very irregular; it is composed of a body and processes. The body, concave anteriorly, presents the *infra orbital foramen;* at the upper and outer part of this is a rough surface to artic. with the malar; *behind* the bone is convex, and articulates with the palate; *internally*, it forms the outer wall of the nasal cavity, and here presents the opening into the antrum, nearly closed by the inf. spongy bone; *inferiorly*, is the *superior alveolar process*, and the palate process, forming the floor of the nose, superiorly, and the roof of the mouth inferiorly. This terminates, on the mesial line, in a thin plate, to receive with its fellow the lower margin of the vomer; it projects in front to form the *anterior nasal spine.* Superiorly is the orbital plate which articulates with the os unguis and ethmoid internally, the palate posteriorly, and the malar externally; behind which is the spheno-maxillary fissure. From the inner and upper part of this bone the *nasal process* runs upwards and inwards to join the frontal bone; its anterior edge is slightly grooved to receive the nasal bone and alar cartilage; the post. is more deeply grooved to form the *nasal canal;* its *internal* surface forms part of the outer wall of the nose, and has attached to it the ant. ext. of the inf. spongy bone. The *body* is hollowed out to form the sinus or *antrum Highmorianum*, which opens into the middle meatus narium.

PALATE. *Sit.* at posterior surface of the sup. maxillary, between it and the pterygoid process of the sphenoid. Irregular, it consists of a horizontal and vertical plate. The *horizontal* or *palate* process, nearly square, is flat and rough below where it forms the post. part of the palate, smooth and concave above to form the floor of the nose; the *posterior* edge is thin, and gives attachment to the velum; its *anterior* is rough and articulates with the superior maxillary; *internally* it artic. with its fellow, and receives in a groove thus formed the vomer.

The *vertical plate* or *nasal process*, ascends from the preceding, and terminates superiorly in two processes; the *ant.* or *orbital* process forms the posterior angle of the floor of the orbit, the *post.* artic. with the body and spongy plate of the sphenoid; the *int. surface* is concave, and forms the outer wall of the nasal cavity; it is divided by a horizontal ridge, which gives attachment to the inf. spongy bone, the part above forms a portion of the middle meatus, that below of the inf. meatus. The *anterior* edge is thin, overlaps the sup. maxillary, and assists in closing the entrance into the antrum; the post. edge joins the pterygoid process.

The *tuberosity*, or *pterygoid process*, proceeds backwards from the lower angle, and is received into the interval between the two pterygoid plates; it is perforated by one or two small holes which lead from the palatine canal. Between the orbital and sphenoidal processes is a fossa, formed into the spheno-palatine foramen by the sphenoid, in which Meckel's ganglion is lodged.

The *inferior spongy bone*, convex internally, concave externally, artic. *behind* with the palate, *in front* with the superior maxillary; from its *upper* margin proceeds upwards a process which joins a descending process from the os unguis to form the nasal duct.

The *os unguis* is placed at the inner and fore part of the orbit, artic. *above* with the frontal, *below* with the inf. spongy bone, *in front* with the sup. maxillary, and *behind* with the orbital plate of the ethmoid; it completes the ethmoid cells anteriorly.

The *nasal bones*. *Sit.* beneath the nasal process of the frontal. *Ant. surface*, convex above, concave below, is perforated by one or two small holes for the escape of small nerves. *Post. surface* concave and grooved by the nasal nerves. *Sup. edge* thick, artic. with the nasal process of frontal and plate of ethmoid. *Inf.* thin, and joins the alar cartilages. *Int.* edge thick, rests against its fellow. *Ext.* groove joins the nasal process of sup. maxill. The nasal process of frontal and nasal lamella of ethmoid support the nasal bones posteriorly.

Vomer. *Sit.* in the lower part of septum narium. *Sup.* margin thick, receives the azygos process of sphenoid. *Ant.* slightly grooved, receives the nasal lamina of ethmoid posteriorly, and triang. cartilage anteriorly. *Post.* separates post. nares. *Inf.* is received in the grooved between sup. maxillary and palate bones; the sides are flat, and covered by pituitary memb.

INFERIOR MAXILLARY is divided into body and rami. *Body*, convex in front, presents on mesial line the *symphysis*, on the side of this is a depression for the depressor labii muscles, to its outer side the *mental foramen*, the termination of the inf. dental canal; concave posteriorly, it presents on the mesial line a chain of projections; to the superior is attached the frænum linguæ, to the middle the genio-hyo-glossi, and to the inferior the genio-hyoid muscle; on each side is a depression for the sublingual gland, and beneath a depression for the digastric.

On the outer surface is an oblique line leading to the coronoid process, to this the platysma and depressor anguli mus. are attached anteriorly, the buccinator posteriorly; on the inner side is a similar line to which the mylo-hyoid and superior constrictor are attached; beneath this is a groove for the mylo-hyoid nerve, and beneath this a depression for the submaxillary gland.

The *rami.* pass upwards and backwards from the body, with which they form the *angle;* on the outer side of this is attached the masseter, on the inner the int. pterygoid; above this is the inferior dental foramen leading to the inf. dental canal, containing the inf. dental vessels and nerves; its lower margin gives attachment to the int. maxillary lig.

The ramus terminates above in two processes, the *coronoid* and *condyloid*, separated by the *sigmoid notch*. The coronoid, pointed, gives attachment to the temporo-maxillary articulation; it is oblong transversely, and is directed backwards and inwards; beneath it is the *neck*, and on the outer side a tubercle for the attachment of the ext. lat. lig.

TEMPORO-MAXILLARY ARTICULATION

is formed by the condyle of the lower jaw inferiorly, the glenoid cavity and transverse foot of the zygoma superiorly, as just described.

Ligaments. Ext. lat. Or. tubercle on outer edge of transverse root of zygoma. *Ins.* outer side of neck of condyle. *Int. lat. Or.* spinous process of sphenoid. *Ins.* lower part of circumference of inf. dental foramen. *Capsular. Or.* Glasserian fissure and circumference of artic. *Ins.* neck of condyle, perforated internally by ext. pterygoid muscle. In the interior of the joint is the *inter-aticular cartilage*, corresponding to the artic. surfaces of the bones; the ext. pterygoid is attached to it internally. The synovial memb. is double, being separated by the int. articular cartilage.

Inter-maxillary lig. thin and weak, descends from the ext. pterygoid process to the root of the coronoid process.

Stylo-maxillary lig. Or. styloid process of temporal bone. *Ins.* inner surface of angle of jaw.

THE SPINE

is composed of twenty-four vertebræ, united by fibro-cartilage and ligaments, and divided into seven *cervical*, twelve *dorsal*, and five *lumbar*. Each vertebra consists of a body, processes, and a spinal foramen.

Cervical vertebræ. Characters—*Body* small, oval transversely, concave above, convex below, anterior margin projects inferiorly. *Transverse process* short and bifid. perforated near its base for the vertebral artery. *Articular process* oblique; *superior* slightly convex, directed upwards and backwards; *inferior* concave, directed forwards and downwards. *Spinous process* short, horizontal, and bifid, the laminæ long and narrow; spinal canal large and triangular.

Peculiar cerv. vertebræ.—First or atlas, no body, or spinous process, composed of two half arches, ant. half arch convex in front, concave behind artic. with the odontoid process; the atlas articulates above with the condyles of the occipital, behind which is a deep groove for the vertebral art.; below with artic. surfaces on the dentata or axis. Transverse process single and projecting. Second. Odontoid process for artic. with preceding, proceeds from upper surface.

Seventh. Spinous process long and projecting posteriorly, and not bifid.

Dorsal vertebræ. Characters—*Body* longer than cervical, heart-shaped, long axis from before backwards, flat above and beneath. On the sides two half artic. surfaces for heads of ribs. *Transverse process* thrown backwards, and marked by an artic. surface on the extremity for the tubercle of the rib. *Articulating process* directed, the superior backwards, the inferior forwards. *Spinous process* long and drooping. *Spinal canal*, small and circular.

Peculiar dorsal vertebræ. First, has one on each side a distinct facette for head of first rib, and a half one for that of second rib, resembles cervical. Tenth, eleventh, and twelfth, one facette for head of rib, eleventh and twelfth no facette, on transverse process for tubercle of rib. Twelfth, inf. artic. process directed outwards.

Lumbar vertebræ. Largest, oval, long axis transverse, flat above and below. *Transverse processes* long, thin, and horizontal. *Artic. processes. Superior*, concave, directed inwards, *inferior* the reverse. *Spinous processes*, short, strong, and square, project posteriorly. *Spinal canal* large and triangular.

Peculiar lumbar vertebra. Last. Transverse process short, inf. artic. process directed forwards. *Body* cut off obliquely to articulate with sacrum.

Ligament of Spine.

Occipito-atlantoidean artic. Anterior and posterior lig. Or. from edge of foramen mag. *Ins.* upper edge of atlas. *Capsular* surrounds artic. surfaces. *Occipito-axoidean lig. Or.* anterior and inner border of foramen mag. *Ins.* posterior surface of body of axis; passes over artic. between first and second vertebræ.

Artic. between atlas and odontoid process.—Beneath preceding are the *transverse* and *oblique* or check ligaments. *Transverse, Or.* a tubercle on each side of oblique process of atlas, passes across the back of odontoid process. *Ins.* opposite side of atlas. *Check, Or.* one on each side from near summit of odontoid process, passes upwards and outwards. *Ins.* inner edge of foramen magnum.

Intervertebral Ligaments.—*Anterior* stretches from the front of the body of the second vertebra, along the anterior surface of the spinal column, down to the extremity of the sacrum; it is expanded, and is attached most firmly to the intervetebral substance. *Posterior* proceeds from the back part of the body of second vertebra, being here continuous with the occipito-axoidean lig. along the post. surface of the bodies of the vertebræ, and is implanted inferiorly into the upper bone of the sacrum; its edges are lunated and well-defined; it adheres most firmly to the inter. vert. substances, opposite to which it expands, but loosely to the bodies of the vertebræ.

Intervertebral substance, or fibro-cartilage.—Between the bodies of the vertebræ (except first and second) to which it firmly adheres, deeper in the lumbar than dorsal, and in these than in the cervical regions, the fibres run obliquely, so as to decussate, and are arranged in concentric laminæ; about the centre, but near the spinal canal, it forms a semi-pulpy substance, of a diamond shape.

Ligamenta subflava extend in pairs, one on each side, from the lamina of the spinous process above to that below, from the axis to the sacrum, so as to close the spinal canal between the laminæ.

Supra spinous lig. extend from post. tubercle on the occipital bone to the sacrum, attached to the extremity of each spinous process, its cervical portion forms the *ligamentum nuchæ*.

Inter. spinous lig. do not exist in the neck, are thin and weak in the back, but are strong in the lumbar region; the fibres stretch from one spinous process to the other.

Inter. transverse lig.—A few irregular fibres, stretching, as their name implies, are only distinct in the lower dorsal and lumbar regions.

The Ribs,

twenty-four in number, are divided into seven true and five false. The former are connected to the sternum by means of the costal cartilages; the latter are only indirectly connected thereto.

Each rib consists of a head, neck, tubercle, and body. *Head*, slightly expanded, divides into two concave facettes, for articulation with the corresponding facettes on the vertebræ, separated by a ridge which gives attachment to the inter-articular lig. *Neck* proceeds outward and backwards, slightly rounded, and terminates in the *tubercle*, a small convex facette, which articulates with the transverse process of each vertebra, beyond which it is rough for the attachment of the posterior costo-transverse lig. *Body*, curved, convex outwards, concave internally, presents posteriorly an oblique line indicating the *angle* of the rib, at which point the bone turns forward. The ridge affords attachment to the sacro-lumbalis tendons, the rest of the body is covered externally by muscles, internally it is lined by the pleura; its *upper* edge is thick and rounded, its *lower* edge slightly grooved for the intercostal vessels; both give attachment to the intercost. muscles. *Ant. extremity*, slightly expanded, is concave, and receives the extremity of the costal cartilage.

The *first rib* is short and flat, and has no angle; its upper surface is marked by two grooves for the subclavian vein and artery, separated by a ridge for the insertion of the scalenus anticus musc.: the head has but one articulating surface for the first vertebra. The tenth, eleventh, and twelfth ribs have but one artic. surface on their heads; the eleventh and twelfth have neither angle nor tubercle; they are short and thin, pointed anteriorly, and loose, hence called *floating* ribs.

The *costal cartilages*, twelve in number, resemble the ribs to which they are attached; the first is short, those of the true ribs articulate with the sternum, those of the false are joined together, and to the cart. of the last true rib, excepting the eleventh and twelfth, which are free.

The *Sternum* is situated on the mesial line of the thorax, in front sloping downwards and forwards. Its *anterior surface*, irregularly flat, is covered by the skin and aponeurosis; its *post. surface*, slightly concave, bounds the ant. mediastinum in front. The sternum is marked laterally by a deep articular surface above for the clavicle, and, beneath this, by articular surfaces, concave, for the cartilages of the true ribs. It consists of two pieces, the superior short and triangular, the inferior long and quadrilateral; to the lower extremity of this is connected the xiphoid cartilage.

Ligaments of the Ribs.—*Anterior costo-vertebral,* stretches from the ant. surface of the head of each rib to be inserted by radiating fibres into the vertebra above and below, and the intervening intervertebral substance. *Inter-articular* runs from the ridge on the head of each rib to the adjoining inter-vertebral substance. *Posterior costo-transverse* lig. Or. extremity of transverse process. *Ins.* non-articular portion of tubercle. *Middle costo-transverse lig.* connects the back part of the rib to the front of the corresponding transverse process. *Inferior costo-transverse,* wanting in the first and last ribs, arise from the lower border of the transverse process, and is inserted into the crest on the upper edge of the rib beneath. The costal cartilages are connected to the sternum by irregular bands of fibres on the ant. and post. surface of this bone: the artic. is lined by synovial membrane.

THE BONES OF THE UPPER EXTREMITY

consist of the clavicle, scapula, humerus, radius, and ulna, the carpal bones, metacarpus, and phalanges.

Clavicle, extends from sternum, upwards and backwards, to the acromion process, curved like an italic *f,* its ant. and post. surfaces are convex and concave in opposite directions. *Sternal end,* thick and triangular, artic. with the sternum. *Body,* nearly cylindrical. *Acromial end,* flattened, and rests on the edge of the acromion process, by a small articulating surface; its *ant. surface* gives attachment to the pect. major and deltoid, its post. to the sterno-mastoid and trapezius, its under surface to the subclavius.

Sterno-clavicular Ligaments.—*Anterior* and *posterior,* stretch from the inner ext. of the clavicle to the sternum. *Inter-clavicular*, extends from one clavicle to the other, behind and a little above the sternum. In the interior of the articulation is an inter-articular cartilage, lined by a double synovial membrane.

Costo-coracoid, Or. cartilage of first rib passes upwards and outwards, enclosing subclavian between two layers. *Ins.* under surface of clavicle and root of coracoid process.

Acromio clavicular ligaments, superior and *inferior*, stretch from one bone to the other.

SCAPULA, *Sit.* upper, lateral, and posterior part of thorax, extending from second to seventh rib. *Int.* or *ant.* surface concave and marked by several ridges, forms the subscapular fossa, and give attachment to the subscap. muscle. *Ext.* or *post. surface* is divided by the *spine* into the supra and infra-spinatus fossæ, muscle; the latter convex, gives attachment to the infra-spinatus, at its lower edge to the teres minor. A quadrilateral flat surface over the angle gives attachment to the teres major. The *spine* commences from the posterior border near its upper third, passes upwards and forwards, and terminates in a quadrilateral flat process, the *Acromion:* to its upper margin is attached the trapezius, to its lower the deltoid. *Superior margin*, thin and short, is notched anteriorly by the supra-scapular notch, in front of which arises the coracoid process, which curves forwards and gives attachment to the coracobrachialis and short head of biceps in front, and the pect. minor internally. *Posterior margin*, or *base*, slightly curved, gives insertion to the rhomboid muscles. *Anterior margin*, thick, gives origin to the long head of the triceps near the neck of the bone. The *superior posterior angle* gives insertion to the levator anguli scapulæ; the *anterior* forms the glenoid cavity, supported on an elongated process, *the neck*. *Glenoid cavity*, oval, long axis directed vertically, is concave, to receive the head of the humerus.

Ligaments of the Scapula. Sup. Proper Lig. crosses the supra scapular notch, and gives origin to the omo-hyoid muscle. *Spino-glenoid* stretches from the under-surface of the spine to the upper and posterior border of the glenoid cavity. *Coraco-acromial* arises broad from the coracoid process, passes upwards, and is inserted narrow into the acromion.

HUMERUS, OR ARM-BONE.—*Head*, rounded and convex, forms little more than half a sphere, is directed upwards and inwards, and joins the glenoid cavity to form the shoulder-joint: it is supported by the *anatomical neck*, forming a deep groove for the attachment of the capsular ligament: beyond this are the *greater* and *lesser tuberosities*, the former, the *external*, has three facettes for the insertion of the supra-spinatus, infra-spinatus, and teres minor muscles. The latter, the *internal*, gives insertion to the subscapularis musc.; between them in front is the bicipital groove through which the long tendon of the biceps runs; beneath them the bone contracts and forms the *surgical neck*. *Body*, or *shaft*, round above, is twisted in the middle, flat and triangular below, its posterior surface gives origin to the second and third head of the triceps, between which is a groove for the musculo-spiral nerve; on the inner surface, near its centre, is a rough surface for the insertion of the coraco-brachialis, and on its outer a triangular rough surface for the insertion of the deltoid; the anterior surface beneath this affords origin to the brachialis anticus muscle.

The lower extremity of the humerus is flattened and twisted a little forwards. From *within outwards* are, the *int. condyle*, projecting, gives origin to the tendon of the pronators and flexors, and the int. lat. ligament; the *epitrochlea*, a projecting eminence, the *trochlea*, an articular surface for the greater sigmoid cavity of the ulna, bounded in front and behind by the ant. and post. humeral fossæ; a slight projection, a rounded small head, the *capitulum*, to artic. with the head of the radius, the *ext. condyle*, slightly projecting for the origin of the ext. lat. lig. and the supinators and extensors; from each condyle a ridge of bones pass es upwards to the humerus, and affords attachment to the intermuscular septa and some of the muscles of the fore-arm.

THE SHOULDER-JOINT

is formed by the glenoid cavity and head of the humerus just described.

Ligaments, Capsulars, arises from neck of scapula, outside brim of glenoid cavity, descends expanding round the head of the humerus, and is inserted into the anatomical neck; strongest inferiorly, weak superiorly, it is perforated in front by the long head of the biceps, internally by the subscapularis tendon, and above by the supra-spinatus.

Coraco-humeral, or *accessory*, extends obliquely downwards and outwards from the coracoid process to the ant. part of the great tuberosity. *Glenoid* surrounds the brim of the glenoid cavity and is continuous with the tendon of the biceps.

BONES OF FORE-ARM.

Ulna forms the inner bone. *Sup. ext.* present greater *sigmoid notch* to artic. with the trochlea of the humerus, bounded above by the *olecranon process*, to which the tendon of the triceps is attached, in front by the *coronoid process* for the insertion of the branchialis anticus; on its outer edge is the *lesser sigmoid notch* for artic. with the side of the head of the radius. *Body*, triangular is divided into three surfaces by three lines, affording attachment to the muscles and intermuscular septa. *Lower ext.* first contracted, then expands and forms a rounded extremity, to artic. with the concavity on the radius, on its inner part is the *styloid process* to which the int. lat. lig. of the wrist is attached.

Radius lies on the outer side of the fore-arm; sup. ext. forms the head, on which is a cup-like cavity to artic. with the capitulum of the humerus; on its side an artic. surface for the lesser sigmoid notch of the ulna; below this is the *neck* and at its lower and inner part the *tubercle* into which the tendon of the biceps is inserted. *Body*, is divided into three surfaces, by three lines for the attachment of muscles and fascia, its *ext. surface* convex presents near its centre a rough surface for the insertion of the pronator teres. *Inf. ext.* expanded and quadrilateral presents in front a concavity for the insertion of the pronator quadratus, *posteriorly* it is marked by three grooves for the passage of the extensor tendons, the middle groove lodges the extensor secundi internodii pollicis tendon; the second, to the ulna side of this, transmits the extensor com. and indicator tendons; the third to the radial side of the first is divided into two for the tendons of the extensores carpi radialis. Along the ext. border is another groove divided into two for the extens. ossis metacarpi and primus internodii pollicis tendons; externally is the styloid process, to which the ext. lat. lig. of the wrist is attached, and on its outer surface the tendon of the sup. longus; internally a concave surface to artic. with the ulna inferiorly, is the artic. surface for the carpus divided partially into two by a slight ridge.

THE ELBOW-JOINT

is formed by the lower ext. of the humerus and upper extremities of the ulna and radius as just described.

Ligaments. Anterior, Or. anterior surface of humerus around ant. humeral fossa, descends obliquely in front of the joint. *Ins.* annular lig. of radius and coronoid process. *Post. lig.* not so well marked. *Or.* around post-humeral fossa. *Ins.* ext. of olecranon process. *Int. lat.* flat and triangular. *Or.* int. condyle of

THE CARPUS.

humerus. *Ins.* side of the olecranon and coronoid process. *Ext. lateral*, *Or.* ext. condyle of humerus. *Ins.* side of annular lig. of radius.

Superior Radio-Ulnar Artic. formed by the side of the head of radius and lesser sigmoid cavity of ulna.

Ligaments. Annular Or. from front and back of lesser sigmoid cavity of ulna, surrounds the edge of the head of the radius. *Oblique lig. Or.* root of coronoid process, descends obliquely outwards. *Ins.* side of radius below its tubercle.

Inferior Radio-Ulnar Articulation, formed by convex side of lower ext. of ulna, and concavity on inner side of radius, the ligaments are *anterior* and *posterior*, composed of irregular fibres attached to each bone close to artic. surfaces.

THE CARPUS consists of eight bones arranged in two rows.

First row, from without inwards as follows, scaphoid, lunar, cuneiform, and pisiform.

Second row, trapezium, trapezoid, magnum, and unciform.

Scaphoid. Largest of first row, convex above, concave beneath, presents four artic. surfaces, *superior* to artic. with the radius, *internal* to artic. with the lunar above, os magnum beneath, *inferior* double to artic. with the trapezium and trapezoides. The rest of the bone is rough for the attachment of ligaments.

Lunar has four articulating surfaces, *sup.* convex to artic. with the radius, *inf.* concave to artic. with the magnum and unciform; *int.* flat to artic. with the cuneiform; *ext.* flat to artic. with the scaphoid.

Cuneiform, wedge-shaped, the base is directed outwards and artic. with the lunar; apex inwards, convex and smooth artic. with the fibro or inter-articular cartilage, *inf.* concave to artic. with the unciform, *ant.* flat to arctic. with the pisiform.

Pisiform, pea-shaped, artic. with but one bone, the cuneiform.

Trapezium, concave above to artic. with **scaphoid**, below, convex from before backwards, concave transversely to artic. with the metacarpal bone of thumb, *int.* artic. with the trapezoid and *inf.* with the second metacarp. bone, *ant.* is grooved by the flexor carpi radialis tendon.

Trapezoid, smallest of this row, artic. above with the scaphoid, *ext.* with the trapezium, *int.* the magnum *inf.* with the second metacarp. bone.

Magnum. The largest of the carpal bones, *sup.* presents a convex head to artic. with the scaphoid and lunar, *inf.* artic. with the second, third, and fourth metacarp. bones, *ext.* with the trapezoid, and *int.* with the unciform, its *post.* surface is large, its *ant.* thin and small.

Unciform, so named from its hook-like processes, which springs from its ant. surface, above its artic. with the lunar, below with the fourth and fifth metacarp. bones, *ext.* with the magnum, and *int.* with the cuneiform. The free surfaces of the carpal bone are rough for the attachment of irregular bands of fibres connecting them together; most of these radiate from the ant. surface of the magnum.

WRIST-JOINT is formed by the lower ext. of the radius above, the scaphoid and lunar beneath, the ulna is separated from the artic. by an inter-articular cartilage.

Ligaments. *Ext. lateral, Or.* styloid process of radius. *Ins.* side of scaphoid, annular lig. and trapezium. *Int. lat. Or.* styloid process of ulna. *Ins.* cuneiform and pisiform bones. *Ant.* and *post.* consist of irregular fibres which descend from the radius to the front and back of the first row of carpal bones. The *anterior and posterior annular ligaments* do not belong to this artic.; they have been described (*see Fasciæ.*) In the interior of this joint is a triangular inter-articular cartilage, between the ext. of the ulna and cuneiform bone, its *apex* is attached in a depression on the outer side of the root of the styloid process of the ulna, its *base* to the right separating the artic. surfaces on the radius. Like the larger joints, this and the intercarpal articulations are lined by synovial membrane.

Metacarpal bones, five in number, resemble each other, *anteriorly* concave and narrow, *posteriorly* broad and convex, *sup. ext.* irregular, artic. with one or more of the carpal bones, *inf. ext.* presents a rounded head, to artic. with the base of the first phalanx. The *first* metacarpal bone is supported by one bone, the trapezium, the *second* by three, trapezium, trapezoid, and magnum, the *third*, by one, os magnum, the *fourth* by two, os magnum and unciform, the *fifth* by one, the unciform. Small artic. surfaces are observed on the sides of the four inner where they touch each other, or one of the carpal bones; they are connected by irregular bands of fibres: that of the thumb possesses lateral ligaments.

Metacarpo-phalangeal articulations are provided with lateral ligaments and a kind of capsular lig.

Phalanges.—Each finger possesses three phalanges, excepting the thumb, which has but two, *anteriorly* concave, posteriorly convex, the metacarp. ext. of the first phalanx is concave, to artic. with the head of the metacarp. bone, the inf. ext. is convex from before backwards, and concave from side to side to artic. with the second phalanx, which presents an opposite arrangement. The last articulation resembles this. The ungual phalanx terminates in a rough pointed extremity. These articulations possess lateral ligaments and a kind of capsular lig. .

THE BONES OF THE INFERIOR EXTREMITIES

Consist of the pelvis, femur, tibia, fibula, tarsus, metatarsus, and phalanges.

The *pelvis* is properly an intermediate structure between the trunk and lower extremities. It is composed of four bones, the sacrum and os coccyx, posteriorly, the ossa innominata on either side.

Os sacrum triangular, the base above the apex below. *Sit.* between ossa innominata, beneath the spine and above os coccyx. *Ant. surface* concave, looks downward and forwards, presents four transverse lines indicating the pieces of which it originally consisted, external to these are the four ant. sacral foramina through which the sacral nerves pass out, between these are smooth surfaces from which the pyriformis arises. *Post. surface*, irregular and convex, presents on mesial line the rudiments of spinous processes, often united into one ridge, beneath which is a triangular channel in which the spinal canal terminates, external to these are grooves for the attachment of muscles, in them are the *post. sacral foramina*, much smaller than the anterior of the exit of post. sacral nerves, external to these is a row of small tubercles analogous to the transverse processes of the vertebræ. The *base* of the sacrum turned upwards and forwards presents an artic. surface for the last lumbar vertebræ, its anterior edge forms the *promontory*, behind it is the opening of the sacral canal, and on each side an oblique artic. process concave, and looking backwards and inwards to form the artic. process of the last lumbar vertebræ. The apex directed forwards and downwards articulates with the first bone of the os coccyx.

Ext. surface presents oblong artic. surface to join the os innominatum, and beneath this, a smooth margin which forms part of the great sacro-sciatic notch.

Os or *ossa Coccygis*—consists usually of three or four irregular small bones, articulated with each other in early life, but consolidated in advanced age.

Os innominatum, irregular in shape, forms the lateral and anterior walls of the pelvis; it is divided into three bones, ilium, ischium and pubes.

Os ilium forms the largest portion of the os innominatum; broad, flat, and triangular, its base is at its upper part, its apex at the acetabulum. *Ext. surface*, turned also a little backwards, concave in front, then convex, and again concave, it presents from above downwards a rough surface, which commences a short distance behind the ant. sup. spinous process, passes backwards between the sup. and inf. semicircular ridges, expanding considerably, and terminates at the post. spinous processes; this gives origin to the gluteus medius muscle, a rough surface at the back part gives origin to the gluteus maximus, beneath is the inf. semicircular ridge, from which and the surface beneath, the gluteus minimus arises; beneath this the bone projects to form the brim of the acetabulum.

Internal surface concave forms the iliac fossa and lodges the iliacus internus musc.: this is bounded inferiorly by the *ilio-pectineal line*, forming the boundary between the true and false pelvis, of which the os ilium forms but little. This line, well marked anteriorly, is indistinct posteriorly: to it the pelvic fascia is attached through its greatest portion, and in front the psoas parvus tendon and Gimbernaut's ligament.

Os ischium forms the lower, outer, and back part of the pelvis. *Ant. surface* turned a little outwards, presents *externally* the *acetabulum*, a deep cup-like cavity for the reception of the head of the femur, with which it forms the hip-joint: this is formed two-fifths and a little more by the ischium, less than two-fifths by the os ilium, and one-fifth by the pubes; it is surrounded by the *brim*, well marked at its upper and back part, shallow elsewhere, and deficient *internally*, where it forms a notch through which vessels and nerves pass to the head of the femur; at the bottom, but nearer the pubes, is an irregular depression which affords origin to the round lig., and lodges Haversian glands. Beneath this is a groove for the obturator ext. tendon, and from this descends a rough ridge which gives origin to the quadratus femoris musc.;

the *anterior* part of the body is thin and presents the *obturator foramen* inferiorly, the *posterior* joins the ilium. and forms part of the great sciatic notch, beneath is the *tuberosity*, from this the ascending *ramus* ascends to join the descending ramus of the pubes.

Os pubis forms the inner and ant. part of the pelvis, from the preceding the *ramus* ascends to terminate in a flat surface attached to its fellow at the symphysis, from the upper extremity of this the *horizontal ramus* passes outwards to join the ilium; its superior margin is the crest, between which and the symphysis is the *angle*. A sharp ridge separates the superior surface of the crest from the posterior, this is the commencement of the ilio-pectineal line; in front of this the ramus is smooth and concave and supports the femoral vessels; at its inner extremity is the *spine* or *tubercle*, to which Poupart's lig. is attached.

The *circumference* of the os innominatum presents, first, the *ant. sup. spinous process* of the ilium; passing backwards, it forms a waving surface, the crest terminates behind in the *posterior superior spine* of the ilium. A little below this, the *posterior inf. spine*, between these, but more internal, an irregular surface to form the sacro-iliac symphysis; passing upwards and forwards the *great sciatic notch*, then the *spine* of the ischium, to which the lesser sciatic lig. is attached, a groove for the exit of the obturator internus tendon, more inferiorly the tuber ischium, then the ascending ramus of this bone, the descending ramus of the pubes, symphysis, horizontal ramus, anterior inferior spine of the ilium, a shallow groove for the exit of the inguino cutaneous nerve, and lastly, the anterior superior spine of the ilium.

Articulations of pelvis.—The last lumbar vertebra is joined to the os sacrum by the inter-vertebral substance and a continuation of the spinal ligaments.

Sacro-iliac symphysis or *synchondrosis*, between the side of the os sacrum and inner edge of os innominatum the bones are connected by anterior and posterior ligaments, and an intervening cartilage, ossified in the old subject.

Great sacro-sciatic ligament extends from the side of the sacrum and coccyx downwards, forwards and outwards, to be inserted into the tuber ischii. *Lesser sacro-sciatic lig.* lies internal to the preceding; smaller, arises from the side of the sacrum and os coccyx, passes outwards, and is inserted, pointed, into the spine of the ischium. The great sacro-sciatic lig. converts the sciatic notch into a foramen, the lesser divides this into two, a superior larger, an inferior smaller, through the former pass out the pyriformis muscle, the gluteal art. and superior gluteal nerve, above the musc.; the greater and lesser sciatic nerves, sciatic and int. pudic art. below the musc. through the lesser sciatic notch, the obturator internus tendon passes out, the int, pudic art. re-enter the pelvis.

Obturator ligament nearly closes the obt. foramen, leaving a small oblique canal at its upper and outer part for the exit of the obturator art. and nerve. The ossa pubis are connected by an *intervening fibro-cartilage*, an *ant. and posterior* and a *sub. pubic* ligament crossing the sub-pubic angle.

The pelvis is divided into the true and false pelvis by the ilio pectineal line: the true pelvis lies above this line, the false beneath it.

FEMUR, or *thigh-bone*, is the only bone in the thigh; it is composed of a head, neck, shaft, and inferior extremity. *Head*, round and hemispherical, presents near its centre, but a little below it, a depression into which the lig. teres is inserted; the *neck* proceeds from this downwards and outwards and joins the shaft at an acute angle; triangular, its *base* joins the shaft, its *apex* the head of the bone, its surfaces are excavated, the inferior is the longest, the superior the shortest; the posterior is longer than the anterior; it is bounded in front and behind by the *ant.* and *post.*

inter-trochanteric lines, extending from one trochanter to the other.

The *shaft* is twisted and curved, convex in front, concave posteriorly. *Sup. ext.* on this are, the *greater* and *lesser trochanters*, the former square, convex externally, forms the highest part of the shaft, on the inner side, near its root is a pit into which the small rotators outwards are attached; the latter, pointed and triangular, lies on its inner and posterior part immediately beneath the neck. *Post. surface* presents on the mesial line the *linea aspera:* this commences above by two lines, from the trochanters, and terminates inferiorly in two lines leading to the condyles, it gives attachment to numerous muscles, and the fascia lata. The *anterior* and *lateral* surfaces of the femur are alternately concave and convex, and are covered by the vasti muscles. *Inf. Ext.* expands to form the *ext.* and *int. condyles*, for the formation of the knee-joint; these are rounded, oblong eminences, the long axis from before backwards, and terminate posteriorly by rounded extremities, they are here separated by a deep notch, in which the popliteal vessels are lodged; *in front* they are partially separated by a concave artic. surface on which the patella glides; the internal projects most inferiorly and posteriorly, the external is wider from side to side, shorter from before backwards, and projects most interiorly. They are incrusted with cartilage and artic. with the tibia below and the patella in front, forming the knee-joint; on the outer surface of each is a projection for the lat. ligaments. Beneath the ext. a groove for the popliteus tendon.

THE HIP-JOINT

is formed by the acetabulum and head of the femur, as just described.

Ligaments. Capsular, Or. round acetabulum ext. to the cotyloid ligament, passes downwards and outwards around the head; *Ins.* neck of femur anteriorly as far as ant. inter-trochanteric line, posteriorly into the centre of the neck Strongest superiorly, and posteriorly, where it receives fibres from the rectus femoris, and anteriorly where it is strengthened by the accessory lig., weakest towards the obt. foramen. *Accessory lig.*, *Or.* ant. inf. spine of ilium passes downwards, inwards, and backwards. *Ins.* lesser trochanter. *Cotyloid lig.* surrounds brim of acetabulum, to which it is attached; it passes across the notch in its inner part and terminates in two extremities attached to its opposite edges. *Structure*, fibro cartilaginous. *Ligamentum teres*, *Or.* bottom of rough surface in acetabulum, passes upwards and outwards, contracting, and is *Ins.* into the depression on the head of the femur. *Transverse.* Is formed by some fibres of the cotyloid lig. crossing the notch in the acetabulum.

Patella. *Sit.* in front of knee-joint, heart-shaped, the *base* above, gives insertion to the rectus tendon, its *apex* below, gives origin to the lig. patellæ, *anterior surface*, convex and rough, is covered by a bursa, *posterior*, covered with cartilage, is divided into two by a vertical ridge, of which the ext. concave corresponds to the ext. condyle, the int. smaller and convex to the int. condyle.

BONES OF LEG. The leg has two bones, *tibia* and *fibula*.

Tibia, extends from knee to ankle-joint. *Sup. ext.* expanded, presents two condyles, concave, for reception of the condyles of the femur, to which they correspond. They are separated by a rough surface in front and behind, also by the spinous process, a projecting eminence. On each side of the head of the tibia is a slight prominence; of these the int. is best marked for the attachment of the int. lat. lig ; on it is a groove for the ant. insertion of the semimembranous tendon; on the under and back part of the outer is a convex artic. surface which joins the head of the fibula ; in front is the tuberosity for the insertion of the lig. patellæ. *Body* triangular, large above, contracts and becomes rounded below, where it again expands to artic. with the astragalus; it is divided into three surfaces by three waving lines, of these the *ant.* is subcutaneous, the others are covered by muscles. *Inf. ext.* somewhat square, presents an *ant. convex surface*, before which descend the extensor tendons, a *post. surface*, grooved for the flexor longus pollicis tendon; *externally* is a rough surface to lodge the lower ext. of fibula, smooth below to artic. with this bone; *internally* is a square projection, the *internal malleolus*, smooth and cartilaginous externally, to artic. with the side of the astragalus, rough and convex internally; its post. edge is grooved for the tibialis posticus and flexor communis tendons. The lower surface, smooth and cartilaginous, is concave to rest on the upper surface of the astragalus.

Fibula, much smaller than the tibia. *Sit.* on its outer side, and, above, a little posterior to it. *Sup. ext.* somewhat rounded, articulates internally with the tibia; rough externally, has a projecting point into which the ext. lat. lig. and tendon of biceps are inserted; below this is the *neck*. *Body*, presents internally a sharp ridge which divides it into two surfaces for the attachment of muscles and fascia. *Inf. ext.* larger than the sup., terminates in a triangular expansion, the *external malleolus;* this is convex internally, and covered with cartilage to artic. with the outer side of the astragalus, rough externally, pointed inferiorly for the attachment of the ext. lat. ligaments, on its posterior surface is a deep groove, for the peronæus longus, and brevis tendons.

THE KNEE-JOINT

is formed by three bones, femur, tibia, and patella, as just described. *Ligaments* divided into external and internal. *External—Ligamentum patellæ. Or.* lower edge of patella, passes downwards and inwards. *Ins.*

lower part of tuberosity of tibia. *Int. lateral* flat and broad. *Or.* back part of inner condyle of femur, descends forwards. *Ins.* inner condyle of tibia, passing down some distance, attached closely to the int. semilunar cartilage. *Ext. lateral*, round and strong. *Or.* tubercle on outer part of ext. condyle of femur, descends forwards. *Ins.* head of fibula. There is sometimes a second ext. lat. smaller, lying posterior to the preceding. *Ligamentum posticum. Or.* inner condyle of tibia, passes upwards and outwards. *Ins.* ext. condyle of femur. Some consider this as one of the insertions of the semi-membranosus muscle.

Internal ligaments. Sit. interior of joint. *Alar* folds of the sigmoid membrane, on each side of the patella. *Ligamentum mucosum. Or.* fatty substance behind lig. patella, passes upwards and backwards. *Ins.* notch between condyle of femur. *Transverse* passes across from the ant. ext. of one semilunar cartilage to that of the other. *Crucial*, the strongest of the internal ligaments. *Anterior* or *external Or.* inner and post part of ext. condyle of femur, descends forwards and inwards. *Ins.* in front of spine of tibia and int. semilunar cartilage. *Posterior* or *internal. Or.* outer and fore part of inner condyle of femur, descends backwards and outwards. *Ins.* behind spine of tibia.

Semilunar cartilages are two interarticular fibro-cartilages, which rest on the upper artic. surface of the tibia, their *extremities* or *cornua*, pointed, are attached in front of and behind spine of tibia; thick and convex *externally*, thin and concave *internally*, the *upper surface* concave, the *under* nearly flat; the inner is the more fixed by the attachments of the int. lateral and ant. crucical ligaments, the ext. is grooved by the popliteus tendon and inf. ext. artic. artery. The knee is lined by the largest synovial memb. of the body, passing up in front of the femur behind the rectus tendon to an extent of three or four inches.

Superior tibio-fibular artic. Ligaments. Anterior and *posterior*, stretching from the tibia downwards and outwards to the head of the fibula.

Inferior tibio-fibular artic. Ligaments. Anterior and *posterior* similar to, but much stronger than, the preceding. *Interosseous*, a strong band of fibres lying between the bones, and connecting them firmly together.

BONES OF FOOT.—Consist of the tarsal and metatarsal and the phalanges.

Tarsal bones are seven in number—viz., *astragalus, calcis* or *calcaneum, navicular, cuboid,* and three *cuneiform;* they may be arranged in two rows, an *anterior* and *posterior*, or an *internal* and *external;* the *posterior* are astragalus and calcis; the *anterior*, navicular, cuboid, and cuneiform; the *external* are calcis and cuboid; the *internal,* astragalus, navicular, and cuneiform.

Astragalus. Sit. upper, inner part, and nearly centre of tarsus; irregular, consists of body, neck and head. *Body,* quadrilateral, presents four artic. surfaces; *superior* square, convex from before backwards, concave transversely to artic. with lower ext. of tibia, continuous internally with an oblong artic. surface for the int. malleolus and externally with a triang. artic. surface for the ext. malleolus. *Inf.* surface presents two artic. surfaces for os calcis, separated by a deep groove; *ext.* larger, broad, and concave; *int.* also *anterior*, and convex; the groove affords attachment to strong interosseous ligaments. *Post. surface* small and non-artic. presents a deep groove for the flexor pollicis longus tendon. Anteriorly is the *neck* which supports the *head*, rounded, convex, articulates with the os navicular in front, and inferiorly rests on calcaneo-navicular lig.

Os calcis.—Largest tarsal bone. *Sit.* inferior, back, outer, and part of inner surface. *Superior surface* presents two artic surfaces for the preceding inferior, to which they correspond, with a similar deep groove; internal to this is a projecting process, the *sustentaculum*, which supports the astragalus, and gives attachment to the calcaneo-cuboid lig. *Inf. surface*, rough and excavated, presents posteriorly, two *tubercles*, the internal larger, for origin of muscles and ligaments. *Int. surface*, concave, transmitting tendons of long flexors and tibialis posticus, post. tibial vessels and nerves. *Ext. surface*, rather flat, presents a tubercle for attachment of ext. lat. lig. of ankle, and separates grooves for the peroneal tendons. *Post. extremity*, convex, projecting behind, presents a rough surface inferiorly for the insertion of the *tendo-Achillis*, above which it is smooth for a bursa-mucosa. *Ant. Ext.* artic. surface for os cuboides.

Navicular or *Scaphoid.*—*Sit.* upper int. and ant. part of tarsus, oval, with long axis directed downwards and inwards. *Post.* presents concave artic. surface for head of astragalus. *Ant.* three artic. surfaces, nearly plane, for three cuneiform bones; its circumference is rough for the attachment of tendons and ligaments. A tubercle on inner and lower part gives attachment to the tibialis posticus tendon. Sometimes its outer surface articulates with the cuboid.

Cuboid. *Sit.* upper, outer, and ant. part of tarsus. *Post. surface*, artic with os calcis; *ant.* with fourth and fifth metatarsal bones; *sup. surface*, flat and rough, for attachment of ligaments; *inf.* irregular and rough, is grooved by a peroneus longus tendon. *Ext.* grooved for peroneus longus tendon. *Int.* presents two artic. surfaces; posterior for scaphoid, ant. for ext. cuneiform.

Cuneiform.—Three in number. *Sit.* int. and ant. part of tarsus. Internal, largest, base below, apex above, artic. *posteriorly* with navicular; *ant.* with first metatarsal bone; *ext.* with second metatarsal and mid. cuneiform; *int.* presents tubercle for insertion of tibialis anticus tendon above, and the posticus beneath. *Middle*, smallest, artic. *behind* with the navicular, *before* with the second metatarsal, and *laterally* with the preceding and ext. cuneiform. *External*, artic. *behind* with navicular, *before* with third metatarsal bone, *int.* with middle cuneiform and second metatarsal, *ext.* with cuboid and fourth metatarsal bone.

Metatarsal bones are five in number. The *first* or internal, the shortest but strongest, the *second* is the largest, they are all convex on the superior or dorsal, concave on the inferior or plantar surface; lateral surfaces are excavated for the lodgment of interossei muscles. *First* is supported behind by the ext. cuneiform, its ant. ext. round and large artic. with first phalanx of great toe, grooved inferiorly for two sesamoid bones, and here gives insertion to the peroneous longus tendon. *Second* is received posteriorly into a sulcus formed by the three cuneiform bones. *Third* rests post. on the ext. cuneiform. *Fourth* and *fifth* on the os cuboides, their ant. extremities rounded and oblique, artic. with the first phalanges of the toes.

Phalanges resemble these of the fingers, except that they are shorter and smaller.

ANKLE-JOINT

Is formed by the lower extremites of the tibia and fibula and upper surface of astragalus, as just described.

Ligaments. *Anterior* and *posterior tibio-tarsal* consist of irregular fibres passing from the tibia downwards to the astragalus in front and behind. *Int. lat. lig.* triangular and flat. *Or.* lower ext. of int. malleolus passes downwards radiating. *Ins.* inner side of calcis, astragalus, and navicular bone. *Ext. lat.* three in number. *Or.* inner surface of ext. malleolus above its point, the *anterior* passes forwards, and is *inserted* into the upper and outer part of astragalus, the *middle* downwards, backwards, and inwards into the os calcis, the *posterior* backwards and inwards into the back part of the astragalus.

Ligaments of Foot.—The only important are the *inf. calcaneo-cuboid* and *inf. calcaneo-navicular:* the former stretches from under surface of os calcis, passes forwards, and is inserted into the cuboid bone, sheath of peroneal tendon, and base of third and fourth metatarsal bones; the latter arises from sustentaculum of os calcis, passes beneath head of astragalus, and is inserted into under part of the navicular bone. Irregular bands of fibres and interosseous ligaments pass between parts of the tarsal bones, but do not merit particular description. The articulations are lined by synovial membrane.

22

CHAPTER X.

ORGANS OF SENSE.

The organs of Sense are five in number—viz., the eye, nose, ear, tongue, and skin.

THE EYE.

The eye is the organ of vision, and is contained in the orbital cavity.

The orbits, or orbital cavities, are two in number. Each orbit forms a quadrangular pyramid, the apex *backwards* at the inner ext. of the foram. lac. orbitale, the base *forwards*, formed by its circumferance. Its four sides or walls are, the *superior* or roof, the *inferior* or floor, the *internal* and *external*. The *roof* is concave, and formed by the orbital plate of the frontal and lesser wing of the sphenoid; the floor is nearly plane, slopes downwards and outwards, and is formed principally by the orbital plate of the superior maxillary, a little anteriorly by the malar, and posteriorly by the palate bone; the *internal* wall, parallel to its fellow, is formed by the nasal process of superior maxillary, os unguis, os planum of ethmoid, and side of the body of the sphenoid; the *external* wall, the shortest thrown outwards, is formed by the sphenoid posteriorly, the malar anteriorly. The *foramina* of the orbit are, the optic, foramen lacerum orbitale, spheno-maxillary, and the anterior and posterior internal orbital. The *circumference* is formed by the frontal above, the malar externally, the malar and superior maxillary inferiorly, and by the nasal process of the superior maxillary internally.

The *palpabræ*, or *eyelids*, cover the eye in front; the superior is the larger and more movable. They are composed of skin, orbicularis palpebrarum, cartilage, fibrous membrane, glands, mucous membrane, and fine areolar tissue.

The *tarsal cartilages* are convex anteriorly, concave posteriorly, one margin is attached by fibrous tissue, the tarsal ligaments, to the circumference of the orbit; the other is free, and contains the eyelashes; the *superior* is broad in the centre, pointed at the extremities; the inferior forms a narrow band.

The *Meibomian glands* or *follicles* lie on the inner surface of each tarsal cartilage, arranged in perpendicular rows, and open on the free margin behind the eyelashes.

The *mucous membrane* lines the interior of the lids, and is a continuation of the conjunctiva, which is reflected over the anterior surface of the eyeball; at the inner canthus it forms a reddish projection, the *caruncula lachrymalis,* and a fold, the *plica semilunaris.*

The *Lachrymal apparatus* consists of the lachrymal gland, ducts, puncta, canals, sac, and nasal duct.

The *Lachrymal gland* is lodged in a deep fossa in the upper and anterior part of the orbit; convex above, concave inferiorly; it is composed of a number of granules united by cellular tissue.

The *Lachrymal ducts* proceed from the gland, are seven in number, and open on the upper and outer surface of the eye by perforating the conjunctiva.

The *puncta lachrymalia* are the small orifices of the lachrymal canals, on the free margin of each tarsal cartilage near its inner extremity; the canals proceed from these in a curved direction, so as to form an elbow shape, and open into the lachrymal sac behind the tendo-oculi.

The lachrymal sac is lodged in the fossa in the inner canthus, and is crossed a little above its centre by the tendo-oculi; it leads into the *nasal duct;* this descends outwards and backwards, and opens into the inferior meatus narium.

The eyeball, or globe of the eye, is nearly spherical, the prominence of the cornea in front giving to its antero-posterior axis a greater length. It is composed of three coats, the sclerotic, choroid, and retina; of three *humours*, the aqueous, lenticular, and vitreous; and three *membranes*, the aqueous, hyaloid, and the aris.

The *sclerotic coat* is the outer fibrous membrane, and forms the posterior four-fifths of the eye; it is dense, firm and resisting; its fibres interlace; it possesses little vascularity, and closely resembles the tunica albuginea testis. It is deficient anteriorly to receive the cornea, and posteriorly is perforated by the optic nerve, forming the *pars cribrosa.*

The *choroid* is the vascular coat of the eye; it lines the interior of the preceding, from which it is separated by the principal ciliary vessels and nerves; its *inner* surface covers the retina; it is perforated posteriorly by the optic nerve, and anteriorly is attached to the sclerotic by the *ciliary ligament,* and externally forms behind the iris a series of folds, the *ciliary processes.* The choroid is of a blackish colour, from the presence of the pigmentum nigrum, and is composed of two coats, an external venous (vasa verticosa,) and an internal arterial (membrana Ruyschiana.)

The *Retina* is the nervous coat, being formed by the expansion of the optic nerve; it lines the interior of the preceding, and rests by its inner surface on the vitreous humour; anteriorly it terminates in the ciliary ligament. The retina consists of three coats, an external cellular or serous (membrana Jacobi;) a central nervous, and an internal vascular, formed by the central art. of the retina. Directly in the axis of vision, that is, about one line and a half external to the entrance of the optic nerve, the retina presents a yellowish spot, *the yellow spot of Sœmmering.* It disappears soon after death.

The *Cornea* completes the globe of the eye in front. Convex *anteriorly*, it is covered by the conjunctiva; its *posterior* surface, concave, is lined by the membrane of the aqueous humour; its edges are bevelled off, and are firmly attached to the margins of the opening in the sclerotic coat. Its structure is horny, and is composed of numerous laminæ, united by fine areolar tissue.

The *Conjunctiva* gives a partial covering to the eye, the anterior portion of which it lines, being reflected upon it from the inner surface of the lids.

The *aqueous humour* fills the anterior chambers of the eye, between the cornea and lens; it is a transparent, thin fluid, and is secreted by the serous membrane lining those chambers, hence called the *membrane of the aqueous humour*. Of the *chambers:* these are separated by the iris, although they communicate through the pupil; the anterior is the larger.

The *Lens* or *lenticular humour*, forms a double convex transparent lens, the posterior convexity being the greater; it is enclosed in a thin membrane forming its *capsule*, and is embedded in a concavity in the forepart of the vitreous humour, to which it is attached by a layer of the hyaloid membrane. Its *structure* is composed of a dense humour, arranged in wedge-like masses, converging towards its centre. Between the capsule and the lens is generally a small quantity of fluid, the liquor Morgagni.

The *vitreous humour* fills the posterior three-fourths of the globe of the eye, and supports the concave surface of the retina; it is intermediate in density between the aqueous and lenticular humour, and is contained in cells formed by its investing membrane, the *hyaloid membrane*. It is traversed by a branch of the central art. of the retina.

The *Iris* is the thin membranous fold which separates the chambers of the eye; it is perforated near its centre to form an opening, the pupil. Its external border is attached to the ciliary ligament, its inner forms the margin of the pupil; its surfaces are lined by the membrane of the aqueous humour, its posterior is blackened by the pigmentum nigrum, and is termed the *uvea*. Its *structure* is composed almost wholly of blood-vessels and nerves: some have assigned to it muscular fibres. The pupil is closed in the fœtus, up to the seventh or eighth month, by the membrana pupillaris.

The *ciliary ligament* is a soft whitish annular band formed by the sclerotic, choroid, retina, and iris; from its internal and posterior part proceed between sixty and seventy triangular folds of the choroid membrane, the ciliary processes.

The eye is supplied with blood from the opthalmic artery; its principal branches are the central art. of the retina, the long, short, and anterior ciliary arteries. Its nerves are the optic and the ciliary, derived from the nasal branch of the ophthalmic of the fifth and the lenticular ganglion.

The muscles of the globe of the eye are the superior, inferior, external, and internal *recti*, and the superior and inferior *oblique*. The recti *arise* from the ligamentary circle surrounding the optic foramen and nerve (the external has a second head from the side of the sella turcica,) and are *inserted* into the sclerotic coat around, and about three lines distant from the cornea, where they form the *tunica adnata*, according to some anatomists.

The *superior oblique* has the same *origin*, but passes upwards, forwards, and inwards, through a pulley in the inner and upper part of the orbit, winds backwards and outwards, and is *Ins.* into the outer and back part of the globe of the eye.

The *inferior oblique* arises from the anterior surface of the floor of the orbit, above the infra orbital foramen; ascends backwards and outwards, and is *Ins.* into the external and posterior part of the globe of the eye, near that of the superior oblique.

By the action of these muscles the eye is turned in every direction; the recti also turn the pupil directly as each acts; the oblique rotate the eye, so as to turn the pupil in opposite directions.

The *levator palpebræ superioris* belongs to the upper lid. *Or.* in common with recti musc. *Ins.* upper edge of tarsal cartilage. *Use*, to raise the upper lid.

The third cerebral nerve supplies all the muscles of the orbit, except the superior oblique, which is supplied by the fourth, and the ext. rectus by the sixth.

THE NOSE.

The nasal cavities are two in number, and are nearly symmetrical, although one is usually larger than the other.

Each *nasal cavity* is pyramidal in shape, the *apex* or *roof* is arched, and is formed by the nasal, frontal, cribriform plate of ethmoid and body of sphenoid. The base or floor nearly plane is formed by the palate-plates of sup. maxillary and palate bones, the inner wall is formed by the septum, composed of the vomer and triangular cartilage; the outer wall, irregular, slopes downwards and outwards, by the ethmoid, os unguis, sup. maxillary, inferior spongy, and palate bones. In it are three *meatuses* narium, the *superior* short and closed anteriorly is formed between the superior and middle spongy bones, the *middle* meatus between the middle and inferior spongy bones, the inferior meatus between the inferior spongy bone and floor; the posterior ethmoid cells and sphenoidal sinus open into the superior meatus; the anterior ethmoid cells and frontal sinus by a common opening, and the antrum by a distinct aperture open into the middle meatus; and the nasal duct into the inferior meatus. The posterior nares open into the pharynx: they are bounded by the vomer internally, the int. pterygoid plate externally, the sphenoid above and the palate plate of the palate bone anteriorly.

The nasal cavities are lined by a fibro-mucous membrane, the pituitary or Schneiderian membrane; it is supplied with blood from the nasal branches of the ophthalmic, labial, and palatine arteries. Its nerves are the olfactory, the nerve of the special sense of smell and filaments from the nasal branch of the ophthalmic and naso-palatine or Meckel's ganglion.

The cartilages of the nose are the *triangular*, which assists in forming the septum narium, the *lateral nasal*, two on each side, *superior* and *inferior*, which form the wings or sides of the nose, the inferior forms the *columna* or ridge between the anterior nares.

THE EAR

is divided into the external, middle, and internal.

The *external ear or auricle* is irregular and is composed of the outer border, the *helix*, anterior and inferior to which is a semicircular prominence, the *antihelix;* in front of the meatus is a triangular projection, the *tragus*, opposite to which is the anti-tragus. The lower pendulous part is the lobe. The depressions are the fossa innominata and the concha. The structure of the external ear is chiefly fibro-cartilage, covered by skin enclosing a few muscular fibres.

Meatus auditorius externus leads from the concha to the membrana tympani which separates it from the middle ear, about one inch or an inch and a quarter in length: it is slightly curved convex above; first runs forwards and then turns a little backwards, its lower wall is longer than the upper; the external half of the meatus is formed of cartilage, deficient anteriorly, the inner is bony. It is lined by a continuation of the cuticle, and contains numerous glands which secrete *cerumen* or wax.

The *middle* ear or *tympanum* lies between the membrana tympani and internal ear.

Membrana tympani is a thin semi-transparent membrane, nearly circular, slightly concave externally, convex internally, which forms a septum between the external and middle ear; to its inner surface is attached the handle of the malleus. Its structure consists of a central fibrous lamina, lined externally by the cuticle, and internally by mucous membrane.

The *Tympanum* forms the middle ear, it is a small cavity of an irregular form; *posteriorly*, it communicates with the mastoid cells, *anteriorly* it leads to the Eustachian tube, *externally* it is bounded by the membrana tympani, and internally by the vestibule. On its inner wall are the *fenestra ovalis*, or vestibuli; below this, the promontory; behind, the *fenestra rotunda* or cochleæ; in front, the canal for the tensor tympani musc.

The *Eustachian tube* is a narrow canal about one inch and a half in length, which runs from the tympanum downwards, forwards, and inwards, and opens in the posterior nares opposite the the inferior spongy bone; its structure is fibro-cartilaginous.

In the anterior inferior wall of the tympanum is the inner opening of the Glasserian fissure.

The tympanum contains four small bones, the *ossicula auditus*, the *malleus*, or hammer, the *incus*, or anvil, the *stapes*, or stirrups, and the os orbiculare.

The *internal ear*, or *labyrinth*, is composed of the cochlea, semicircular canals and vestibule.

The *cochlea* forms the anterior part of the labyrinth, of a conical form, its apex is directed downwards, forwards, and outwards, its *base* forms the bottom of the meatus auditorius internus.

The *semicircular canals* are three bony tubes in the petrous bone, behind the vestibule, with which they communicate by five openings: they are named the *superior, posterior, and inferior* or *horizontal*.

The *Vestibule* forms the central portion of the labyrinth; it is a small space between the fenestra ovalis and meatus auditorius internus, and presents seven large foramina, viz. fenestra ovalis, fenestra rotunda, and the five orifices of the semicircular canals, besides smaller openings of the cribriform plate and the *aqueductus* vestibuli.

The portio mollis of the seventh nerve supplies the internal ear, on which it is expanded, as the retina is in the interior of the eye; it is the acoustic nerve. The tympanum is traversed by the portio dura, chorda tympani, and some smaller nerves. (See Nervous System.)

THE TONGUE

is the organ of taste; its situation and shape need not be described. Its structure consists of a peculiar erectile tissue, supposed by some to contain muscular fibres, to which are added the proper lingual muscles already described; it is lined throughout its greatest extent by mucous membrane, which contains numerous follicles, glands, and papillæ scattered over its surface. The papillæ are divided into three classes the calyciform, fungiform, and conical, or filiform.

The *calyciform* are arranged on the dorsum of the tongue, near its base is a V shape; at the apex, which is turned backwards, is the *foramen cæcum*. The *fungiform* are chiefly towards the tip and sides. The *conical* or filiform occupy the greater part of the dorsum.

Tue tongue is supplied with blood by the lingual artery, its nerves are the gustatory and lingual, the former being the nerve of special sense, the latter that of motion.

THE SKIN

is the organ of the special sense of touch, as well as the seat of peculiar sensibility. It consists of three layers, the epidermis or cuticle, rete mucosum, and durmis or cutis vera.

The *epidermis* or *cuticle* forms the external layer; it is composed of small laminæ or plates, slightly overlapping and connected to the cutis vera by numerous fine filaments, small excretory ducts and the sheaths of the hairs.

The *rete mucosum*, although described as a distinct layer of membrane, is considered by some to be granules of colouring matter deposited in the interstices of the cells of the cuticle and cutis, so as to give to the skin its various colours in the different races.

The *dermis* or *curtis vara* is composed of condensed areolar tissue, numerous blood-vessels, nerves, and exhalent vessels; its deep surface is loose and irregular, and is connected to the subjacent parts by cellular tissues, vessels, &c., its free surface is provided with numerous projections forming the *papillæ*, in which the sense and sensibility of the skin reside. The dermis contains numerous glands—viz., the sudoriferous, or sweat glands, the sebaceous, Meibomian, ceruminosæ, &c.

The *hairs* and *nails* are appendages of the skin, and are modifications of the cuticle.

THE END.

www.ingramcontent.com/pod-product-compliance
Lightning Source LLC
Chambersburg PA
CBHW032205230426
43672CB00011B/2517